33rd Annual North American Membrane Society Annual Meeting (NAMS 2024)

Membranes for Decarbonization

Santa Fe, New Mexico, USA
11 - 15 May 2024

ISBN: 979-8-3313-0056-2

Printed from e-media with permission by:

Curran Associates, Inc.
57 Morehouse Lane
Red Hook, NY 12571

Some format issues inherent in the e-media version may also appear in this print version.

North American Membrane Society (NAMS)
University of Arkansas
Ralph E. Martin Department of Chemical Engineering
3202 Bell Engineering Center
Fayetteville, AR 72701-1201
USA

Phone: +1 479 575-7926
Fax: +1 479 575-7926

names@uark.edu

Additional copies of this publication are available from:

Curran Associates, Inc.
57 Morehouse Lane
Red Hook, NY 12571 USA
Phone: 845-758-0400
Fax: 845-758-2633
Email: curran@proceedings.com
Web: www.proceedings.com

NORTH AMERICAN MEMBRANE SOCIETY

33rd ANNUAL MEETING

2024 MAY 11TH – 15TH

SANTA FE COMMUNITY CONVENTION CENTER, SANTA FE, NM

MEMBRANES FOR DECARBONIZATION

NAMS 2024 – The North American Membrane Society

Program Book

Diamond Level

MEWBOURNE COLLEGE OF EARTH AND ENERGY
The UNIVERSITY of OKLAHOMA

INSTITUTE FOR RESILIENT ENVIRONMENTAL AND ENERGY SYSTEMS
The UNIVERSITY of OKLAHOMA

NSF

Gold Level

MilliporeSigma

PALL

Silver Level

DOW

the human energy company Chevron

emi twente

AIR PRODUCTS

ZwitterCo

OSMOSES

PROMETHEUS CARBON NEUTRAL E-FUELS

cytiva

Bronze Level

Air Liquide

MAXWELL ROBOTICS

MTR

Aqua Membranes
PRINTED SPACER TECHNOLOGY®

STERLITECH Corporation

Gale & DB Bhattacharyya

VEOLIA

Exhibitors

NSF IUCRC • Membrane Science, Engineering & Technology Center MAST

HIDEN ISOCHEMA

Surface Measurement Systems
World Leader in Sorption Science

Friends of NAMS

GT Georgia Tech

GT Georgia Tech College of Engineering Separations Science and Engineering Center

Happy Hour

BASF
We create chemistry

Conference chairs

Mahdi Malmali
Michele Galizia
Nitish Mittal
Benny D. Freeman

WIFI: CONVENTION CENTER
Password: no password

Table of Contents

Welcome to NAMS 2024

Fellow NAMS Members,

I want to personally welcome you to the 2024 NAMS annual meeting.
The organizers have put together an incredible program for all the members to enjoy. As you know, the NAMS meeting is NO VACATION!! Weekend workshops are given by the best professors in the field, six sessions in parallel, social hours, evening poster sessions, and the Banquet!
We hope that during that busy week, you will also enjoy the fabulous historic city of Santa Fe.
As I mentioned in the previous Membrane Quarterly, the year was quite busy, and the board of directors worked hard to put together multiple services for the members, such as the student mentoring program, online webinars, award preparedness meetings, etc. Keep an eye on what is available to the members in the members section of our new website.
Remember that, due to the change in secretary this year, all communications will be sent via nams@membranes.org.
Now, just make sure you stay hydrated (Santa Fe is at an elevation of 7,199 feet - third highest city in America), get ready to learn a lot and have fun!!!

Best regards,
Marie-Eve Langevin
NAMS President
President@membranes.com

Message from the 2024 NAMS Conference Chairs

It is our pleasure to welcome you to Santa Fe for the 33rd Annual Meeting of the North American Membrane Society. It has been our honor to serve as co-chairs for this year's NAMS meeting.

We will use Whova as a platform for program information, message distribution, and networking, so please make sure you download the app on your phone and bookmark it on your computer. We expect to have 150 posters and 200 talks on a wide range of membrane-related topics. These contributions are of the highest quality, which will make NAMS 2024 a very engaging conference.

We thank our dedicated oral and poster session chairs, and the NAMS Board. We would like to acknowledge especially our sponsors who have provided substantial funds to support the conference and related events, such as the Membrane Workshops and the traditional Lunch with Legends.

We are truly excited to host you in Santa Fe and look forward to seeing many old friends and meeting new ones. We hope you have a great time at NAMS 2024!

Mahdi Malmali, Texas Tech University, Lubbock
Michele Galizia, University of Oklahoma, Norman
Nitish Mittal, Exxon Mobil, Houston
Benny D. Freeman, The University of Texas at Austin, Austin
NAMS 2024 Conference Co-Chairs

NAMS 2024 Anti-Harassment Policy

NAMS 2024 is dedicated to providing a harassment-free conference experience for everyone. We will not tolerate harassment of conference participants (including sexual harassment and assault) in any form. Conference participants violating these rules may be sanctioned or expelled from the conference without a refund at the discretion of the conference organizers. Harassment includes but is not limited to: comments related to gender, gender identity, and expression, sexual orientation, physical appearance, race, or religion; unwanted sexual attention; unwanted physical contact; threats of violence; intimidation; and stalking.

Reporting: If someone makes you or anyone else feel unsafe or unwelcome, please report it as soon as possible to one of the conference co-chairs. If you are being harassed by someone, notice that someone else is being harassed, or have any other concerns, please contact one the conference co-chairs through the Whova App or in person.

Enforcement: If a participant engages in harassing behavior, event organizers retain the right to take any actions to keep the event a welcoming environment for all participants. This includes warning the offender or expulsion from the conference with no refund. Event organizers may take action to redress anything designed to, or with the clear impact of, disrupting the event or creating a hostile environment for attendees. We expect participants to follow these rules at all event venues and event-related social activities.

NAMS 2024 EDUCATION INNOVATION FELLOWSHIP

This award supports innovators in membrane education, and educational initiatives that introduce membranes into the science curriculum at colleges and universities.

Stephanie G. Wettstein

Dr. Stephanie G. Wettstein is an Associate Professor in the Chemical and Biological Engineering department at **Montana State University** (MSU) where she joined the department in 2012. She received her B.S. in Paper Science at the University of Wisconsin (UW)– Stevens Point and worked as a Process Engineer for three years at Kimberly-Clark after graduation. Dr. Wettstein then obtained her Ph.D. from the University of Colorado in the area of zeolite membrane separations under Profs. John Falconer and Rich Noble and then completed a two-year post-doc at the UW – Madison in the area of catalysis and separations extractions under the guidance of Prof. James Dumesic. Her technical research is a combination of her past experience and is focused on catalysis and separations to produce products from renewable resources. Dr. Wettstein has over 12 years of experience teaching mass transfer unit operations including the junior-level separations course and the senior-level unit operations laboratory. In both classes, she has incorporated industry-based experiences from her time at Kimberly-Clark including redesigning the senior lab course to make it more industrially relevant.

Heather C.S. Chenette

Dr. Heather Chenette is an Associate Professor of Chemical Engineering at **Rose-Hulman Institute of Technology**. Her professional interests include leveraging qualitative methods to understand and enhance student learning in the classroom and creating opportunities for students to learn about polymers, membrane materials, and bioseparation processes through research experiences. She graduated with a B.S. in Engineering (Harvey Mudd College) and worked as a chemical engineer for a NanoH2O, Inc., developing and testing membranes for desalination before earning her PhD in Chemical Engineering from Clemson University, developing membranes for chromatographic bioprocessing applications while also obtaining a Certificate in Engineering and Science Education. She has been on the faculty at Rose-Hulman since 2014. Heather attended her first NAMS Annual meeting in 2010 and presented at the NAMS Student Workshop in 2015. The NAMS Education Innovation Fellowship supported project will assess the current state of membrane education in higher education, identify areas of greatest need of educational resources, and develop and promote appropriate educational resources. She is also an active member of ASEE and AIChE, where she currently serves as a Director for the Separations Division, and regularly presents scholarly work and engages in the Education Division.

NAMS Fellows are recognized members of NAMS who have demonstrated ongoing excellence in membrane engineering and practice. Election as Fellow is in recognition of both "service to NAMS" and "highly significant professional accomplishment in the membrane field."

John Pellegrino

Dr. John Pellegrino spent 16 years on the NAMS Board of Directors between 1997 and 2023, during which he was NAMS President in 2004; co-founded the Alan S. Michaels Award and founded the Elias Klein Travel Award. He also served and variously chaired the Finance, Awards, Workshops, Annual Meeting and Fellows Committees. In 1995, he created the first NAMS website (www.membranes.org). John also co-chaired two NAMS Annual Meetings (Boulder in 2000 & Estes Park, CO in 2021) and founded and co-instructed the Measurement Methods for Membranes workshop that has been often presented at NAMS and ICOM meetings between 1995-2024. Recently, he has been helping to produce the Member and Membrane News feature for the Membrane Quarterly.

John is currently a Research Professor in the Department of Mechanical Engineering at the University of Colorado-Boulder. Prior to joining the research faculty at CU, he was vice-president of engineering for Santa Fe Science and Technology, Inc., a defense contract research company trying to commercialize polyaniline hollow fibers. He also spent 14 years as a member of the National Institute of Standards and Technology (NIST) where he led the membrane science and technology program and co-chaired and managed the NIST–Advanced Technology Program Focused Competition on Selective Membrane Platforms. He also spent 7 years as a chemical engineer with duPont and Rohm & Haas. His career technical highlights include recently co-developing submicron surface patterning of membranes. This work has stimulated a broad swath of current academic and industrial researchers. He was also co-PI on a team providing the first experimental evidence for molecular sieving through single angstrom-sized pores in the 2-d carbon allotrope, graphene. This work has been cited ~1200 times. He was also a co-PI on team providing the first systematic study of how membrane systems should be used to effectively to remove natural organic matter from municipal water treatment systems and minimize impacts on water's taste, odor, and potential disinfection byproducts. This work has been cited ~1150 times. His earliest efforts was as co-PI on a team providing the first systematic study of structure-property relationships in mass transfer through perfluorosulfonated ionomer (Nafion®) which was emerging as the "gold standard" in polymer electrolyte membranes.

Uwe Beuscher

Dr. Uwe Beuscher is a Technical Leader at W.L. Gore & Associates, Inc., the world leader in fluoropolymer materials for a wide variety of applications. Dr. Beuscher received his Diplom-Engineer degree in Mechanical Engineering from RWTH Aachen (Germany) and his Ph.D. in Chemical Engineering from Clemson University. After joining W.L. Gore & Associates, Dr. Beuscher contributed for over 25 years to developments in a wide variety of membrane and separation applications including polymer fuel cells, gas-liquid contactors, barrier materials, gas separation, liquid purification, advanced microfiltration, adsorption, catalysis, and chromatography. Currently, Dr. Beuscher is developing new product ideas and applications for improved sustainability and environmental remediation. He is also leading the Modeling and Simulation Team for the Performance Solutions Division, which develops multiscale mathematical models and simulation/analysis tools to understand and optimize the performance of Gore's materials, products, and manufacturing processes. Dr. Beuscher has authored or co-authored more than thirty technical papers and book chapters and has presented frequently at international meetings. He was a member of the NAMS board of directors for 8 years, chaired the Awards committee during that time, and served as the NAMS president in 2015. Dr. Beuscher co-chaired the 2021 NAMS annual meeting in Estes Park, CO, and has co-taught the Measurement Methods for Membranes for the last several years. He is a founding member of the World Association of Membrane Societies (WA-MS) and has been active on its Steering Committee since its inception in 2017.

Winston Ho

Dr. W.S. Winston Ho is a Distinguished Professor of Engineering in the William G. Lowrie Department of Chemical and Biomolecular Engineering and the Department of Materials Science and Engineering at The Ohio State University. Winston has been a member of NAMS since its formation. He served as Director of NAMS, 2005 – 2011. Currently, he has served as Separation Engineering Section Editor, Current Opinion in Chemical Engineering and on several advisory boards including the Journal of Membrane Science. He has chaired many membrane and separations sessions at NAMS and AIChE meetings. Moreover, he gave an invited plenary lecture at the NAMS 2016 Annual Meeting, and he has developed a membrane course and taught it every year since 1999 for educating next-generation membranologists and wide-spreading membrane knowledge. Before teaching for 25 years in 2024, he had 28 years of industrial R&D experience in membranes and separation processes. He was elected to the National Academy of Engineering, USA in 2002 in recognition of his distinguished contributions to engineering including the commercialization of gas treating technologies and the development of supported liquid membrane technology and membrane separation models. A New Jersey Inventor of the Year (1991), Winston holds more than 60 U.S. patents, generally with foreign counterparts, in membranes and separation processes. He is an AIChE Fellow and has received several AIChE awards.

Student Fellowship Award

The NAMS Student Fellowship Awards are presented annually to outstanding graduate students in the area of membrane science and technology.

David Kitto works with Professor Jovan Kamcev as a 5th year Ph.D. candidate in the Department of Chemical Engineering at the **University of Michigan**. He performed his bachelor studies in Chemical Engineering and Chemistry at the University of Minnesota - Twin Cities. David's research interests include the modeling, characterization, design, and preparation of polymeric ion-exchange membranes for both water and energy applications. His thesis work has integrated a comprehensive approach to membrane synthesis with large data sets to realize high-performing new materials possessing uniquely high charge densities. His work has been recently recognized with awards by WA-MS at ICOM 2023 and by the ACS Division of Polymer Chemistry at ACS-Spring 2024.

Jianan Gao is a Ph.D. candidate working with Dr. Wen Zhang in the Department of Civil and Environmental Engineering at the **New Jersey Institute of Technology**. He received an M.S. degree in Municipal Engineering and a B.S. degree in Water Supply and Drainage Science and Engineering from Qingdao University of Technology, China. His research focused on developing an innovative electrochemical membrane technology for wastewater treatment. His work achieved remarkable efficiency in nitrate reduction and ammonia recovery and gained over 900 citations. His work was recognized with the 2024 NAMS Student Fellowship Awards, the 2023 CAPEES Founding President Best Paper Award, the 2023 Kenneth S. Stoller Award, the 2023 The C. Ellen Gonter Environmental Chemistry Award, the 2022 Heh-Won Chang PhD Fellowship in Green Chemistry Award, and the 2021 Graduate Student Award in ACS Division of Environmental Chemistry (ENVR).

Matthew Webb is a 4th year Ph.D. candidate in the Department of Sustainable, Chemical, Biological and Materials Engineering at the **University of Oklahoma**, where he is co-advised by Dr. Michele Galizia and Dr. Sepideh Razavi. Matthew received his Bachelor of Science in Chemical Engineering from the University of Oklahoma in 2020, with special distinction. His primary research focuses are currently geared towards the development of next generation polymer-based gas separation membrane materials exhibiting enhanced selectivity and stability. Matthew has received the NAMS Elias Klein Founders' Award in 2023, and has co-authored several scholarly papers. His thesis work revolves around the investigation and optimization of the design and characterization of selective and stable facilitated transport films, for applications such as natural gas purification or olefin/paraffin separation.

Young Membrane Scientist Award

The NAMS Young Membrane Scientist Award is for outstanding individuals who are starting their professional careers in membrane science and technology (≤5 yr from terminal degree).

Dr. Kecheng Guan is an Assistant Professor of the Research Center for Membrane and Film Technology at **Kobe University**, Japan. He obtained his Ph.D. in Chemical Engineering from Nanjing Tech University in 2019 and joined Kobe University as a postdoctoral researcher in the group of Prof. Hideto Matsuyama that same year. In 2020, he was promoted to Assistant Professor. Kecheng's research primarily involves developing membrane materials and processes for water treatment, organic solvent separation, and resource recovery. He is also interested in investigating selective transport within confined nanopores. Currently, he leads two research projects funded by the Japan Society for the Promotion of Science (JSPS) Grant-in-Aid for Early-Career Scientists and the Sumitomo Foundation Basic Science Research Projects, with a main focus on developing fouling-resistant polymeric membranes for seawater distillation.

Dr. Joshua Moon is an Assistant Professor of Chemical Engineering at the **University of Florida**. Prior to joining UF faculty in 2023, he worked as a postdoctoral researcher at the University of California, Santa Barbara under the guidance of Profs. Craig Hawker and Rachel Segalman as part of the Center for Materials for Water and Energy Systems (M-WET). He received his Ph.D. in Chemical Engineering in 2019 from the University of Texas at Austin where he was an NSF Graduate Research Fellow under the advisement of Profs. Benny Freeman and Donald Paul. His research group aims to tackle Grand Challenges in clean water, energy, and sustainability using chemically-modular design of polymer membranes and adsorbents coupled with a fundamental understanding of molecular transport in polymers informed by polymer physics and thermodynamics. Current applications for his research span across carbon capture, hydrogen and hydrocarbon separations, environmental remediation of water resources, and sustainable use of polymer materials.

Dr. Samuel Lounder is a Senior Research Scientist at **ZwitterCo, Inc.** He earned his Ph.D. in Chemical and Biological Engineering in 2022 from Tufts University under the guidance of Professor Ayse Asatekin. His Ph.D. and startup research involve the design/synthesis of functional polymer chemistries to develop self-assembled amphiphilic membranes for various water treatment applications. His work has been recognized by awards such as the Tufts Institute of the Environment (TIE) research fellowship, an International Filtration News (IFN) student spotlight feature, and a first-place prize in the NAMS poster competition. Prior to his graduate studies, Sam received a B.S. in Chemical Engineering from the University of Maine.

Permeance Prize

The Permeance Prize is for outstanding individual who are in the middle of their professional careers in membrane science and technology (≤15 yr from terminal degree). The purpose of the Award is to recognize and encourage individual contributions in the field of membrane science and technology, and service to NAMS by a mid-career member of the Society.

Jason Bara

Jason Bara is a Professor at The University of Alabama in the Department of Chemical & Biological Engineering. Jason received a B.S. in Chemical Engineering from Virginia Commonwealth University and a Ph.D. in Chemical Engineering from the University of Colorado at Boulder. He has authored more than 160 peer-reviewed research publications in the areas of gas separations, ionic liquids and green solvents, and polymer membranes. In recent years his research interests have expanded to additive manufacturing, the depolymerization and upcycling of plastic wastes, and liquid-liquid extractions for recovery of critical elements. To date, Jason has been awarded 15 patents for discoveries in these areas. In recognition of his research, he has received awards including the Early Career Fellow from the Industrial & Engineering Chemistry Division of ACS (2021), and the FRI/John G. Kunesh Award from the Separations Division of AIChE (2015).

Jeffrey McCutcheon

Jeffrey McCutcheon is the Centennial Term Professor in the Chemical & Biomolecular Engineering Department at the University of Connecticut. He received a B.S. in Chemical Engineering from the University of Dayton and his Ph.D. in Chemical Engineering from Yale University. For nearly 20 years, he has pioneered work in membrane-based separations, notably in the areas of osmotic processes and membrane formation. He has published over 100 refereed publications and has several patents on membrane technology and has served the separations community as a Director for both the AIChE Separations Division and the North American Membrane Society (NAMS) and recently served as President of NAMS. He currently serves as the Deputy Topic Area lead for Materials & Manufacturing Topic Area in the National Alliance for Water Innovation, the Department of Energy's $100M "Water Hub" dedicated to supporting the development of desalination technology in the United States. He has received numerous awards including the 3M Nontenured Faculty Award, the Solvay Advanced Polymers Young Faculty Award, The DuPont Young Faculty Award, and the FRI/John G. Kunesh Award from the AIChE Separations Division. He was the winner of the 2019 Global Water Summit Water Technology Idol competition for his work on 3D printed membranes and was inducted into the Connecticut Academy of Science and Engineering in 2021. In 2017, he was named the Executive Director of Fraunhofer USA Center for Energy Innovation and served for 3 years before taking the Center to its now independent status as the Connecticut Center for Applied Separations Technologies (CCAST).

Elias Klein Travel Supplement Award

Supports students needing limited financial assistance. This award program is named in honor of Elias Klein, whose vision and spirit guided the founding of NAMS in 1985.

Amir Hossein Mostafavi	University of Arkansas
Amos Taiswa	Montana Technological University
Caleb Song	University of Colorado Boulder
Daniel Gardeno	University of Chem & Techn in Prague
Guangyu Zhu	New Jersey Institute of Technology
Jasneet Kaur Pala	The University of Alabama
Joelle Scott	University of Washington
Karen Guillen-Cuevas	Clemson University
Km Prottoy Shariar Piash	West Virginia University
Lianna Johnson	Case Western Reserve University
Mara Leach	University of Kentucky
Omid Jazani	Clarkson University
Pablo Dean	Massachusetts Institute of Technology
Parisa Taheri	University of Houston
Rajan Jain	University of Colorado Boulder
Riccardo Lovison	University of Colorado Boulder
Sandeep Aryal	Oakland University
Sulaimn Dhamer	University of Kentucky
Utkarsh Misra	Indian Institute of Technology Bombay
Wang (Tony) Hongchen	University of Toronto

NSF Student Travel Supplement Program Recipients

Arjun DeJesus	University of Connecticut
Aron griffin	Colorado School of Mines
Fathy Attia	The State University of New York at Buffalo
Hammed Balogun	Georgia Institute of Technology
Hasib Mahbub	Texas Tech University
Isabella Le	University of Alabama
Jose Edgardo Lopez Cazares	Arizona State University
Josef Schneider	University of Chemistry and Technology, Prague
Juan Zhai	Texas Tech University
Kate Kimball	The University of Texas at Austin
Lucas Condes	The University of Oklahoma
Lucca Madeo Cortarelli	University of Kentucky
Martina Jagielski	University of Connecticut
Matthew Webb	University of Oklahoma
Missoury Wolff	Clemson University
Moein Hajian	Texas Tech University
Moses Adejumo	Clarkson University
Mostafa Tabatabaei	University of Oklahoma
Nerissa Kiprotich	University of Kentucky
Sean Bannon	University of Virginia
Seancarlos Gonzalez	University of Washington
Vraj Shah	University of Alabama
Will Box	University of Oklahoma
Yi-Hung Lin	Auburn University

NAMS
The North American Membrane Society

➢PROGRAM-AT-A-GLANCE
NAMS 2024 Meeting

Date	Time (*MST)	Event	ROOM
Saturday May 11	8:00 AM – 5:00 PM	Membranes for Water Treatment & Reuse	Coronado/DeVargas
		Measurement Methods for Membranes	Peralta/Lamy
Sunday May 12	8:00 AM – 5:00 PM	Membrane Gas Separations	Coronado/DeVargas
		Membranes in Bioprocessing	Peralta/Lamy
	2: 00 PM – 5:00 PM	Student Workshop	O'Keeffe/Milagro/Kearny
	7:30 AM – 6:00 PM	Registration	SFCCC Lobby
	6:00 PM – 9:00 PM	Welcome Reception	SFCCC Courtyard
Monday May 13	7:00 AM – 5:00 PM	Registration	SFCCC Lobby
	8:00 AM – 8:30 AM	Coffee Break	SFCCC Lobby/Corridor
	8:30 AM – 10:50 AM	Parallel Technical Sessions I	Breakout rooms
	11:00 AM – 12:00 PM	Plenary Lecture I	Sweeney E-F
	12:00 PM – 1:45 PM	Lunch with Legends	Hilton S. Fe Historic Plaza Ballroom
	12:00 PM – 1:45 PM	Lunch Break	-
	1:45 PM – 3:45 PM	Parallel Technical Sessions II	Breakout rooms
	3:45 PM – 4:00 PM	Coffee Break	SFCCC Lobby/Corridor
	4:00 PM – 6:00 PM	Parallel Technical Sessions III	Breakout rooms
	7:30 PM – 10:00 PM	Poster Session	Sweeney A-C
Tuesday May 14	7:00 AM – 5:00 PM	Registration	
	8:00 AM – 8:30 AM	Coffee Break	SFCCC Lobby/Corridor
	8:30 AM – 10:50 AM	Parallel Technical Sessions IV	Breakout rooms
	11:00 AM – 12:00 PM	Plenary Lecture II	Sweeney E-F
	12:10 PM – 1:00 PM	Lunch Break	-
	1:00 PM – 3:00 PM	Parallel Technical Sessions V	Breakout rooms
	3:00 PM – 3:30 PM	Coffee Break	SFCCC Lobby/Corridor
	3:30 PM – 5:30 PM	Parallel Technical Session VI	Breakout rooms
	5:30 PM – 6:30 PM	NAMS Business Meeting	Nambe (SFCCC upstairs)
	6:30 PM – 7:30 PM	Happy Hour- Closing and Award	La Fonda Hotel Ballroom
	7:30 PM – 11:00 PM	Banquet Dinner	La Fonda Hotel Ballroom
Wednesday May 15	7:00 AM – 5:00 PM	Registration	
	8:00 AM – 8:30 AM	Coffee Break	SFCCC Lobby/Corridor
	8:30 AM – 10:50 AM	Parallel Technical Sessions VII	Breakout rooms
	11:00 AM – 12:00 PM	Plenary Lecture III	Sweeney E-F
	12:00 PM – 1:00 PM	Lunch Break	-
	1:00 PM – 3:00 PM	Parallel Technical Sessions VIII	Breakout rooms

*MT – US Mountain time, UK +7 hrs, Germany +8 hrs, Saudi Arabia (in Riyadh) +9 hrs, Inda (IST) +11.5 hrs, Singapore +14 hrs, China +14 hrs, South Korea +15 hrs, Japan +15 hrs.

NAMS
The North American Membrane Society

➢PROGRAM-AT-A-GLANCE
NAMS 2024 Meeting

Date	Time (MST)	Event	Room
Monday May 13	8:30 AM – 10:50 AM (Parallel oral sessions)	**Electrochemical separations**	Coronado/DeVargas
		Advanced membranes for gas and vapor separation I	Milagro/Kearny
		Functional membr. fabrication and environmental sustainability	O'Keeffe
		Process intensification and decarbonization	Peralta/Lamy
	11:00 AM – 12:00 PM	**Plenary Lecture I**	Sweeney E-F
	12:00 PM – 1:45 PM	**Lunch with Legends**	Hilton S.Fe
	1:45 PM – 3:45 PM (Parallel Oral Sessions)	**Water treatment and desalination processes**	Coronado/DeVargas
		Adv. membr. for gas and vapor separation II	Milagro/Kearny
		Hollow fiber membranes	O'Keeffe
		Liquid extraction and contaminant separation	Peralta/Lamy
	3:45 PM – 4:00 PM	**Coffee Break**	SFCCC Lobby
	4:00 PM – 6:00 PM (Parallel oral sessions)	**Desalination and osmotic processes I**	Coronado/DeVargas
		Novel synthetic strategies for gas separation	Milagro/Kearny
		Tailored membrane formation	O'Keeffe
		Lithium separation	Peralta/Lamy
	7:30 PM – 10:00 PM	**Poster Session**	Sweeney A-C
Tuesday May 14	8:30 AM – 10:50 AM (Parallel oral sessions)	**Membranes for carbon capture I**	Milagro/Kearny
		Membrane fouling	Coronado/DeVargas
		Organic molecule and solute selective separations	Peralta/Lamy
		Membrane distillation and pervaporation	O'Keeffe
	11:00 AM – 12:00 PM	**Plenary Lecture II**	Sweeney E-F
	12:00 PM – 1:00 PM	**Lunch Break**	
	1:00 PM – 3:00 PM (Parallel Oral Sessions)	**Desalination and osmotic processes II**	Coronado/DeVargas
		NAMS awards session	O'Keeffe
		Advanced understanding and synthetic efforts in solute selective sep.	Peralta/Lamy
		Biochemical applications	Milagro/Kearny
	3: 00 PM – 3:30 PM	**Coffee Break**	SFCCC Lobby
	3:30 PM – 5:30 PM (Parallel Oral Sessions)	**Desalination and osmotic processes III**	Coronado/DeVargas
		NAMS keynote session I	Peralta/Lamy
		2D membrane materials and processes	Milagro/Kearny
		Viral and antibody filtration	O'Keeffe
	5:30 PM – 6:30 PM	**NAMS business meeting**	Nambe
	6:30 PM – 7:30 PM	**Happy Hour + Closing and Awards**	La Fonda Hotel Ballroom
	7:30 PM –10:00 PM	**Banquet dinner**	La Fonda Hotel Ballr
Wednesday May 15	8:30 AM – 10:50 AM (Parallel oral sessions)	**Membranes for carbon capture II**	Milagro/Kearny
		Specific ion separations	Peralta/Lamy
		Scaling up: bridging the gap between labs and industry	O'Keeffe
		Contaminants removal from water I	Coronado/DeVargas
	11: 00 AM – 12:00 PM	**Plenary Lecture III**	Sweeney E-F
	12:00 PM – 1:00 PM	**Lunch Break**	
	1:00 PM – 3:00 PM (Parallel Oral Sessions)	**Mixed matrix and composite membr. for gas separation**	Milagro/Kearny
		Solute selective separation modeling	O'Keeffe
		NAMS keynote session II	Peralta/Lamy
		Contaminants removal from water II	Coronado/DeVargas

FEDERAL PL.
Santa Fe
COMMUNITY CONVENTION CENTER
LOADING DOCK
GRANT AVE.
SWEENEY F
SWEENEY E
SWEENEY C
SWEENEY D
SWEENEY A
SWEENEY B
O'KEEFFE
MILAGRO
KEARNY
CORONADO
DeVARGAS
PERALTA
LAMY
COURTYARD
VISITOR INFORMATION CENTER
LOBBY
COMMUNITY GALLERY
YOU ARE HERE
ARTS COMMISSION
CONVENTION SERVICES
W. MARCY ST.
TERRACE
COURTYARD
POJOAQUE
NAMBE
OHKAY OWINGEH
TESUQUE
FIRST LEVEL
SECOND LEVEL
SANTA FE
COMMUNITY CONVENTION CENTER
201 W. Marcy St.
Santa Fe, New Mexico 87501
PRIMARY EXIT
SECONDARY EXIT
ACCESSIBLE EXIT
FIRE EXTINGUISHER
MEETING SPACE
EXTERIOR SPACE
OFFICE SPACE
ACCESSIBLE RAMP
VISITOR INFO CENTER
RESTROOM

TERRA Conference
June 7 - 11, 2022 (6/6 set up)
1st Floor Breakouts

*SFCCC projector/screens
*Alliance sound

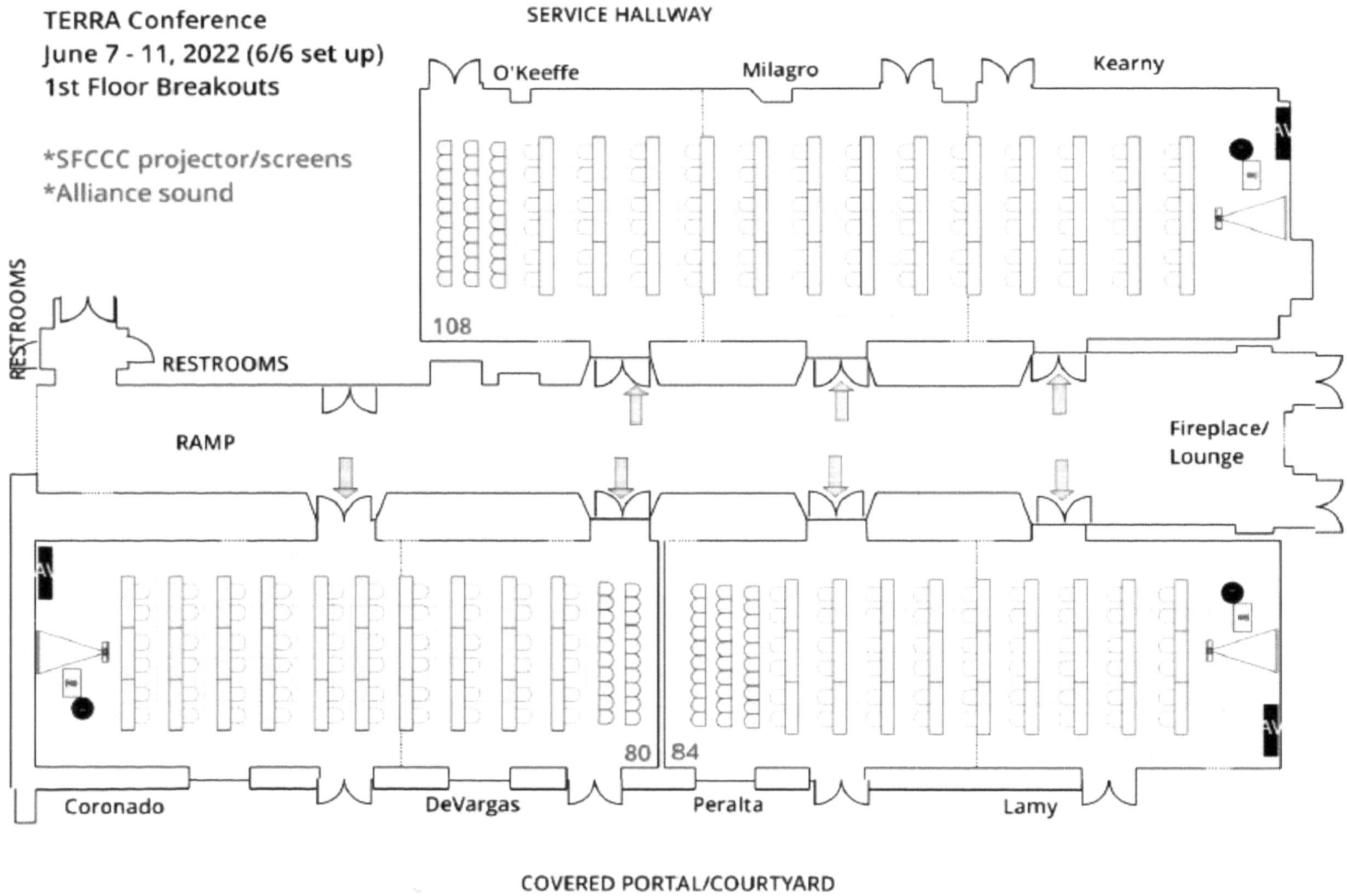

Parallel Sessions	Room: Coronado/DeVargas Electrochemical separations	Room: Milagro-Kearny Advanced membranes for gas and vapor separation I	Room: O'Keeffe Functional membrane fabrication and environmental sustainability	Room: Peralta-Lamy Process intensification and decarbonization
Chairs	Caleb Funk caleb.funk@dupont.com Abhishek Narayan Mondal abhishek.mondal@donaldson.com	Nitesh Bhuwania nbhuwania@chevron.com Chen Zhang czhang71@umd.edu Simona Liguori sliguori@clarkson.edu	Margaret Lee margaret_lee@pall.com Priyesh Wagh priyesh.wagh@gmail.com	Albert Kim albertsk@hawaii.edu Syed Islam islamsz@ornl.gov
8:30 AM	Oral 1 - Understanding Polyfluoroalkyl Substance Selectivity in Anion Exchange Membrane Systems Sam Thompson (Univ. of Kentucky)	Oral 8 - Enhanced H2 transport in polybenzimidazole by blending with a highly permeable polyimide Narjes Esmaeili (Univ. of Buffalo)	Oral 15 - Scaled Manufacturing of Supported Loose-NF Membranes with Eco-Friendly Solvents for Water Treatment Applications David Lu (Univ. of Kentucky)	Oral 22 - Zeolite Membrane Reactor for CO2 Hydrogenation to Methanol: A Simulation Study on Comparison of Water and Methanol Selective Membranes Jerry Lin (ASU)
8:50 AM	Oral 2 – Nanofiltration membranes with thermo- and pH-responsive behavior from lyotropic liquid crystal templating Reza Foudazi (University of Oklahoma)	Oral 9- Tightening ultramicropores of structurally engineered carbon molecular sieve membranes for precise gas separation Zhongyun Liu (GaTech)	Oral 16 - Sustainable additive manufacturing of polysulfone membranes for liquid separations Brian Leonard (West Virginia University)	Oral 23 - Catalytic Membrane Reactors for combined separation and contaminant degradation Charles F. De Lannoy (McMaster Univ.)
9:10 AM	Oral 3 - Charge-patterned nanofiltration membranes with immobilized polystyrene sulfonate particles and polyethylene imine Ali Benhboudi (PSU)	Oral 10 - Computer vision-based measurement of pure- and mixed-gas swelling in polymer membranes exhibiting configurational free volume Will Box (Univ. of Oklahoma)	Oral 17 - An alternative application of 2D carbon-based membranes for efficient air-separation and H2 purification from gas mixtures Dániel Gardenö (University of Chemistry and Technology in Prague)	Oral 24- Computational Fluid Dynamics and Systems Engineering to minimize membrane-based carbon capture costs Alejandro Pedrozo (CMU)
9:30 AM	Oral 4 – Treating Acidic Metal Wastewaters with Ceramic Ion Exchange Membranes Emily Rabe (Membrion)	Oral 11 - Hydrogen- and acetylene-stable silver ion facilitated transport membranes: the effect of salt content on fractional free volume and gas permeability Matt Davenport (UT Austin)	Oral 18 - Understanding the mechanism of additive-assisted isoporous membrane formation through X-ray scattering Noah Wamble (UT Austin)	Oral 25 - Innovating Membranes for Carbon Capture: From Material Synthesis to Process Prototype Yang Han (OSU)
9:50 AM	Oral 5 - Synthesis and Testing of Electrochemical Activity in a Polyether Sulfone-Based Ultrafiltration Membrane for Contaminant Removal Utkarsh Misra (IITB, India)	Oral 12 - Exploring the separation efficiency of ultra-selective 3D CANAL hydrocarbon polymeric membranes for hydrogen-based separations after long-term aging Jing Ying Yeo (MIT)	Oral 19 - Elucidating the Structure-Property-Performance Relationship of Incorporating Waste Polymethyl Methacrylate in Polyvinylidene Fluoride Membrane Fabrication Joseph Danner (Univ. of Alabama)	Oral 26 - Advancements of Responsive and Catalytic Membrane Reactor for Water and Halo-Organics Remediation Dibakar Bhattacharyya (Univ. of Kentucky)
10:10 AM	Oral 6 - Electroconductive Filters and Membranes for Desalination and Wastewater Treatment Swatantra Pratap Singh (IITB, India)	Oral 13 - Evaluation of air addition into methane dry reforming reaction in Pd/YSZ membrane reactor for syngas production Omid Jazani (Clarkson)	Oral 20 - Enhanced bipolar membrane water disassociation reaction with layer-by-layer assembled clay composite junctions Nadia Boulif (TUE)	Oral 27 - Intensified propane dehydrogenation in carbon molecular sieve hollow fiber membrane reactors Chen Zhang (Univ. of Maryland)
10:30 AM	Oral 7 - Performance of Commercial Bipolar Membranes for Ocean Alkalinity Enhancement Application Charles Francois de Lannoy (McMaster University)	Oral 14-Synthesis and Characterization of ZnCl2-Immobilized Molten Salt (IMS) membrane for NH3 separation Moses Adejumo (Clarkson)	Oral 21- Advancing sustainable membrane fabrication by investigating the effect of biodegradable and non-biodegradable solvents on polylactic acid (PLA) membranes Isabel Escobar (Univ. of Kentucky)	Oral 28- Mathematical Modeling & Optimization of a Water-Gas Shift Membrane Reactor for Low-Cost H2 Production with CO2 Capture Hani Hawa (Precision Combustion)

Monday Afternoon

Parallel Sessions	**Room: Coronado/DeVargas** Water treatment and desalination processes	**Room: Milagro-Kearny** Advanced materials for gas and vapor separation II	**Room: O'Keeffe** Hollow fiber membranes	**Room: Peralta-Lamy** Liquid extraction and contaminant separation
Chairs	Ngai Yin Yip n.y.yip@columbia.edu Yian Chen yian.chen@nrel.gov	Nitesh Bhuwania nbhuwania@chevron.com Chen Zhang czhang71@umd.edu Simona Liguori sliguori@clarkson.edu	Stephen DeWitt stephen.dewitt@arkema.com Jianjia Yu jianjia.yu@nmt.edu	Yoram Cohen profyc@gmail.com Isabel Escobar Isabel.escobar@uky.edu
1:45 PM	Oral 29 - Revisiting the value of high-flux seawater reverse osmosis membranes with variable-flow energy recovery Sangsuk Lee (UCLA)	Oral 35- What Membranes Do We Need for Industrial Propylene/Propane Separations? Yang Liu (MTR)	Oral 41 - Experimental and theoretical evaluation of feed flow collar design for shell fed hollow fiber membrane modules James Tran (UToledo)	Oral 47- XpressKleen 1 nm expanded PTFE membrane for harsh chemical filtration Zhefei Li (PALL)
2:05 PM	Oral 30 - Performance and Environmental Evaluation of Biochar-based Membranes Synthesized from Various Petroleum- and Bio-derived Solvents Abelline Fionah (Univ. of Kentucky)	Oral 36 - Scale-up of perfluoro(butenyl vinyl ether) (PBVE) and perfluoro(2,2-dimethyl-1,3-dioxole) (PDD) copolymers for the separation of R-410A Abby Harders (Kansas U)	Oral 42 - Development of a modular device for evaluating hollow fiber membrane permeance Nathan Mullins (McMaster Univ.)	Oral 48 - 2,3-Butanediol Separations using Membrane Assisted Liquid-Liquid Extraction Shailesh Singh Dangwal (ORNL)
2:25 PM	Oral 31 - High Recovery Desalination of High Mineral Scaling Source Water via Integrated RO-NF Concentrator, Concentrate Desupersaturation and Second Stage RO-NF Desalting Yoram Cohen (UCLA)	Oral 37 - Selective photonic gasification of strained oxygen clusters on graphene for overcoming the trade-off between pore size and density Luc Bondaz (EPFL)	Oral 43 - Physical Aging of Defect-Free Asymmetric Functionalized PIM Hollow Fiber Membranes Philippe Jean Baptiste (MIT)	Oral 49 - Continuous membrane demulsification for in situ solvent extraction of carboxylic acid Yian Chen (ORNL)
2:45 PM	Oral 32 - Electrifying membrane distillations: Localized Joule heating of laser-induced graphene membranes Md Hasib Al Mahbub (Texas Tech)	Oral 38 – Pall Filtration is Enabling the Energy Transition Soondeuk Jeung (PALL)	Oral 44 - Improved nanofiltration through inverted membrane structures Tjerk R Watt (UTwente)	Oral 50 - Membrane nanoprecipitation enables the continuous production of uniform nanoparticles at higher concentrations of the polymer solution than with a batch system Lidietta Giorno (ITM-CNR)
3:05 PM	Oral 33 - The analytic solution of interfacial concentration with observed rejection ratio during dead-end membrane filtration Albert Kim (Hawaii Univ.)	Oral 39 - New class of imidazolium-based ionenes for gas separation membranes Jason Bara (Univ. of Alabama)	Oral 45- Tannic acid-coated hollow fiber for dehydration: From material to membrane and module Lakshmeesha Upadhyaya (KAUST)	Oral 51 - High temperature ceramic membrane separations for carbon circularity: From lube oil purification to chemical recycling of plastics Pieter Vandezande (VITO)
3:25 PM	Oral 34- High-Throughput Experimentation for High-Pressure Reverse-Osmosis Membranes Hongchen Wang (Univ. of Toronto)	Oral 40 - Adjustable microstructure design of a polyphthalonitrile precursor and the derived novel carbon molecular sieve membrane for gas separation Chun Po Hu (NTU)	Oral 46 - Developing a Thin Film Composite Facilitated Transport Hollow Fiber Membrane for Post Combustion CO2 capture Oguz Karvan (EMI Twente BV)	Oral 52 – Targeted Separation Scheme of Polyurethane Depolymerization Products Taysha Talenar (ASU)

Monday Evening

Parallel Sessions	**Room: Coronado/DeVargas** Desalination and osmotic processes I	**Room: Milagro-Kearny** Novel synthetic strategies for gas separation	**Room: O'Keeffe** Tailored membrane formation	**Room: Peralta-Lamy** Lithium separation
Chairs	Jovan Kamcev jkamcev@umich.edu Jonathan Maisonneuve maisonneuve@oakland.edu Muskan Sonker muskan.s@gatech.edu	Victor Kusuma victor.kusuma@netl.doe.gov Ali Rownaghi ali.rownaghi@netl.doe.gov Jaason Williams p.j.williams@shell.com	Lakshmeesha Upadhyaya lakshmeesha.upadhyaya@kaust.edu.sa Luis Francisco Villalobos lf.villalobos@usc.edu Varun Vakharia Varun_vakharia@pall.com	Ngoc Bui ngoctbui21@ou.edu Harekrushna Behera harekrushna.behera@austin.utexas.edu
4:00 PM	Oral 53 - Compaction of high-pressure water treatment membranes: real-time quantification and support layer contribution Weijian Ding (USC)	Oral 59- **Permeance Prize: Utilizing Molecular Design to Integrate Ionic Liquids within High Performance Polymer Gas Separation Membranes: Unprecedented Opportunities for Control over Material Properties and Performance** **Jason Bara (Univ. of Alabama)**	Oral 65 - Sulfonated Polyamide Thin Film Composite Membranes for Electrodialysis Deniz Ipekci (UCONN)	Oral 71- Co-Ion Specific Effect on Li Salt Sorption in Cation Exchange Membranes Rachel Huang (UT Austin)
4:20 PM	Oral 54 - Membrane heterogeneity alters our interpretation of effective energy barriers to transport Nathanael S Schwindt (UC Boulder)	Oral 60 - Few-Cycle Atomic Layer Deposition to Nanoengineer Polymeric Materials for Efficient Gas Separations Leiqing Hu (Univ. at Buffalo)	Oral 66 - Compaction-resistant Thin Film Crosslinked Composite (TFX) Ultra-high Pressure Reverse Osmosis Jishan Wu (UCLA)	Oral 72 - Mixed counter-ion sorption in ion-exchange membranes: understanding the role of membrane water content Harsh Patel (UMich)
4:40 PM	Oral 55 - Evidence for solution-diffusion in pressure-induced transport of solvent through dense polymer membranes Kevin Reimund (UT Austin)	Oral 61 - Amine Modified PIM-1 for Highly Selective Olefin/Paraffin Separation Bo Wei Cynthia Chen (Univ. Of Notre Dame)	Oral 67 - CFD modeling of high-flux plate-and-frame membrane modules for industrial carbon capture Cheick Dosso (CMU)	Oral 73 - Preparation of lithium-ion selective mixed matrix membrane for lithium recovery from real brine solution via diffusion dialysis process Sandip Pal (UCLA)
5:00 PM	Oral 56 - Is the Role of Concentration Polarization Overstressed in Membrane Filtration Processes? Lianfa Song (Texas Teach)	Oral 62 – A New Microcapillary Film Carbon Molecular Sieve Membrane Jay Liu (DOW)	Oral 68 - Free Energy Surface Landscape of Interfacial Polymerization for Polyamide Membrane Formation Xianghong Qian (Univ. of Arkansas)	Oral 74 - Mixed binary alkali halide salt transport in PEGDA Aubrey Quigley (UT Austin)
5:20 PM	Oral 57 - Modeling and experimental analysis of pH influence in boron separation in reverse osmosis Fernan Martinez (KAUST)	Oral 63 - The Effect of Crosslinker Concentration on Competitive Release Kinetics and Physical Characteristics of Thermo-Responsive, Lignin-Based Soft Composites Missoury Wolff (Clemson)	Oral 69- The influence of additives on polystyrene membranes created by temperature-induced phase separation Sulaiman Dhameri (Univ. of Kentuky)	Oral 75 - Tuning ionic selectivity in nanoporous graphene membranes using atomic layer deposition and polyelectrolyte assembly Simar K Mattewal (MIT)
5:40 PM	Oral 58 - Flexible RO-NF System Configuration Operation in the Mode of Partial Concentrate Recycle for Inland Water Desalination Yakubu Jarma (UCLA)	Oral 64 –	Oral 70 - The dynamics of Interfacial Polymerization: insights on film stability during formation Guy Z Ramon (Technion)	Oral 76 - Lithium Selective Artificial Ion Channels for Lithium Separation Harekrushna Behera (UT Austin)

Tuesday Morning

Parallel Sessions	Room: Milagro-Kearny Membranes for Carbon Capture I	Room: Coronado/DeVargas Membrane fouling	Room: Peralta-Lamy Organic molecules and solute selective separations	Room: O'Keeffe Membrane distillation and pervaporation
Chairs	David Hopkinson david.hopkinson@netl.doe.gov Joshua D. Moon Joshua.moon@ufl.edu Zhongyun Liu zhongyunliu@gatech.edu	Nima Shahkaramipour nima.shahkaramipour@pentair.com Manish Kumar manish.kumar@utexas.edu	Geoff Geise geise@virginia.edu Ryan Lively ryan.lively@chbe.gatech.edu	Amos Taiswa ataiswa@mtech.edu Allyson McGaughey allysonmcgaughey@unm.edu Lucy-Mar Camacho lucy.camacho@tamuk.edu
8:30 AM	Oral 77- Leveraging competitive sorption in microporous polymer membranes to enhance CO2- and H2S-specific gas separation performance Pablo Dean (MIT)	Oral 84 - Optimizing electrospun composite-fiber patterns on nanofiltration membranes for improved biofouling resistance Amos Taiswa (Mich Tech)	Oral 91 - Organic Solvent Reverse Osmosis Separations of Hydrocarbon-Alcohol Mixtures using Matrimid Membranes Bennett Marshall (Exxon Mobil)	Oral 98 - Long-term Operation of Pilot-scale Membrane Bioreactor – Membrane Distillation for Potable Water Reuse Luke Presson (Univ. of Arizona)
8:50 AM	Oral 78 - Amine Functionalized Supported Ionic Liquid Membranes (SILMs) for CO2/N2 Separation Antoine Chamoun-Farah (UT Austin)	Oral 85 - Effect of Pattern Size Relative to Oil Droplet Size on Oil Fouling in Nanofiltration Lauren Ward (Univ. of Alabama)	Oral 92 - Fluorine-Rich Poly(Arylene Amine) Membranes for the Separation of Sustainable Fuels and Hydrocarbons Yi Ren (Gatech)	Oral 99 - Combined chlorine dioxide–membrane distillation for the treatment of produced water Mahdi Malmali (Texas Tech)
9:10 AM	Oral 79 - Effect of Viscosity of Deep Eutectic Solvent on CO2 Capture Performance in an Energy Efficient Membrane Contactor Based Process Syed Islam (ORNL)	Oral 86 - The effect of alternating electric fields on membrane colloidal fouling Guy Ramon (Technion)	Oral 93 - Organic solvent nanofiltration and reverse osmosis performance of carboxylated PIM-1 in neat and inter-crosslinked mixed membrane configurations Lucas Condes (Univ. of Oklahoma)	Oral 100 - Membrane Distillation and Crystallization for Treating Hydraulic Fracturing Produced Water Ranil Wickramasinghe (Univ. of Arkansas)
9:30 AM	Oral 80 - Incorporating phenyl groups in polyorganosilica membranes to improve hydrothermal Stability for H2/CO2 Separation Vinh Bui (Univ. at Buffalo)	Oral 87 - Optimization of Fouling Resistant Polyelectrolyte Coacervate Membranes via Aqueous Phase Separation Shao-Hsiang Hung (UMASS)	Oral 94 - Volatile fatty acid selectivity during electrodialysis: role of solution composition on volatile fatty acid partitioning and mobility in an anion exchange membrane Holly M Haflich (UNC)	Oral 101 - Advances in the Characterization and Digital Reconstruction of 2-D and 3-D Pore Structure Jay Werber (Univ. of Toronto)
9:50 AM	Oral 81 - Effect of Film Thickness and Temperature on the CO2 Sorption in Poly(ethyleneimine) Christopher Stafford (NIST)	Oral 88- Mechanism of crystallization and membrane scaling mitigation via carbon nanotube-containing spacer in membrane process Seongeom Jeong (Univ. of Pusan)	Oral 95 - Membrane-based Ultrapurification of Isopropyl Alcohol Kamalesh Sirkar (NJIT)	Oral 102 - Fast and selective desalination using pressure-driven distillation with nanoscale air gaps Omar Laris (UC Boulder)
10:10 AM	Oral 82 - 100% Carbon Capture from a Carbon Dioxide-Argon Flue Gas using Polaris Membrane Pingjiao Hao (MTR)	Oral 89 - Atomic layer processes for membrane fabrication and surface modification Tamar Segal-Peretz (Technion)	Oral 96 - Co-Transport of Alcohols and Carboxylate Ions in Sulfonated Ion Exchange Membranes with PEGPEA as a Blocking Group Pravin Parasakthi Aravindhan (Auburn)	Oral 103 - From surface wetting to internal wetting: wetting front progression in membrane distillation Shounak Joshi (Univ. of Southern California)
10:30 AM	Oral 83 - Thin Film Composite Membranes from Polymers of Intrinsic Microporosity for CO2 Separation Sunshine Iguodala (Imperial College)	Oral 90 - Fabrication of polymer brush grafted anti-biofouling ultra-filtration membrane using atom transfer radical polymerization Swatantra P Singh (IIT Bombay)	Oral 97 - Fundamental investigation about the effect of chemical structure of foulant on fouling behavior of polyamide membrane in organic solution Atsushi Matsuoka (Kobe Univ.)	Oral 104 –

Tuesday Afternoon

Parallel Sessions	Room: Coronado/DeVargas Desalination and osmotic processes II	Room: O'Keeffe NAMS Awards	Room: Peralta-Lamy Advanced understanding and synthetic efforts in solute selective separations	Room: Milagro-Kearny Biochemical applications
Chairs	Jovan Kamcev jkamcev@umich.edu Jonathan Maisonneuve maisonneuve@oakland.edu Muskan Sonker muskan.s@gatech.edu	Xiaobo Dong xiaobo.dong@cytiva.com Michele Galizia mgalizia@ou.edu	DB Bhattacharyya db@uky.edu Shashank Kamdar Skamdar@dow.com	Hee Jeung Oh hjoh@psu.edu Cristiana Boi cboi@ncsu.edu
1:00 PM	Oral 105 - Biomimetic membranes incorporating artificial water channels for high-performance water reverse osmosis desalination Mihail Barboiu (Univ. of Montpellier)	**Oral 111 - How click chemistry can enable design and investigation of facilitated CO2 separation membranes** **Joshua D. Moon (UFlorida)**	Oral 117 - Microwave dielectric relaxation spectroscopy: A technique to inform ion transport in hydrated polymer membranes Geoffrey Geise (UVA)	~~Oral 123~~ - Membrane Optimization and Filtration Strategies Using Nanoparticles for Enhanced Efficiency in Gene Therapy Applications Mara Leach (Univ. of Kentucky)
1:20 PM	Oral 106 - Upcycled PVC Support Layer from Waste PVC Pipe for Thin Film Composite Nanofiltration Membranes Atta Ur Razzaq (Univ. of Alabama)	**Oral 112 - Rational design, synthesis, and characterization of facilitated transport membranes exhibiting enhanced permeability, selectivity, and stability** **Matthew Webb (Univ. of Oklahoma)**	Oral 118 - Insights into methods of determining selectivity of porous ultrafiltration membranes Suzana Ivandic (UT Austin)	Oral 124 - Combined Hydrocyclone-Tangential Flow Microfiltration for Bioreactor Harvesting Amir Hossein Mostafavi (Univ. of Arkansas)
1:40 PM	Oral 107 - Maximizing the Wetting Resistance of Fluorine-free Omniphobic Membranes for Hypersaline Wastewater Desalination Yinchuan Yang (UBC)	**Oral 113 - Surface coatings for fouling-resistant membrane distillation** **Kecheng Guan (Kobe Univ.)**	Oral 119 - Mixing and Matching: Amphiphilic Polyampholytes and Polyelectrolyte Complexes that Mimic Zwitterionic Behavior for Membrane Applications Ayse Asatekin (Tufts)	Oral 125 - Charged biosponge polymers for capturing toxic chemotherapy drugs to minimize chemotherapy Hee Jeung Oh (Pen State University)
2:00 PM	Oral 108 - Direct Membrane Osmometry for Rapid Osmotic Pressure Measurements at High Molality Mi Zhang (UCONN)	**Oral 114 - Ultrahigh Charge Density Ion-Exchange Membranes** **David Kitto (UMich)**	Oral 120 - Highly rigid and microporous polyimine membranes by interfacial polymerization for efficient hydrocarbon separations Taehoon Lee (MIT)	Oral 126 - Mechanisms of Flux Decline in Crossflow Ultrafiltration of Whey Protein: Comparing Different Models to Experimental Data Burcu Ozdemir (UCLA)
2:20 PM	Oral 109 – XPRIZE Water Scarcity Hossein Atoufi (XPRIZE Water Scarcity)	**Oral 115 -Tuning the selectivity of fouling resistant zwitterionic copolymer membranes** **Sam Lounder (ZwitterCo)**	Oral 121 - Real-Time, Non-Destructive Monitoring of Supported Liquid Membranes for Biochemical Recovery Using Electrochemical Impedance Spectroscopy Hiroki Fukuda (UBC)	Oral 127 - Protocol for purification of circular RNA from a self-splicing reaction by ultrafiltration Karen Cuevas (Clemson)
2:40 PM	Oral 110 - Delamination Mitigation of Modified Bilayer Membranes using Single Slot Die Roll-to-Roll Manufacturing Process Minwoo Jung (GaTech)	**Oral 116 - Upcycling wastewater nitrate into ammonia fertilizer via electrified membrane** **Jianan Gao (NJIT)**	Oral 122 - Surface Modification of Nanofiltration Membranes for Selective Nutrient Recovery Km Prottoy S. Piash (WVU)	Oral 128 - Patterned anion exchange membrane for improved performance of glucose fuel cells Thomas A Dos Santos (Cuanschutz)

Tuesday Evening

Parallel Sessions	Room: Coronado/DeVargas Desalination and osmotic processes III	Room: Peralta-Lamy NAMS Keynote I TALKS in this session have a 30 min slot	Room: Milagro-Kearny 2D membrane materials and processes	Room: O'Keeffe Viral and antibody filtration
Chairs	Jovan Kamcev jkamcev@umich.edu Jonathan Maisonneuve maisonneuve@oakland.edu Muskan Sonker muskan.s@gatech.edu	Mahdi Malmali mahdi.malmali@ttu.edu Grigorios Panagakos gpanagak@andrew.cmu.edu Shouliang Yi shouliang.yi@hotmail.com	Reza Foudazi rfoudazi@ou.edu Ying Labreche ying_labreche@pall.com	Hee Jeung Oh hjoh@psu.edu Cristiana Boi cboi@ncsu.edu
3:30 PM	Oral 129 - **Permeance Prize: Thin Film Composite Membranes are not Just for Reverse Osmosis Anymore: Recoupling Materials Discovery with Thin-Film Manufacturability** **Jeff McCutcheon (UCONN)**	**3:30 - 4:00 PM** **Oral 135 - A Tale of Two Carbon-Capture Properties: Bulk and Thin Films of Mixed Matrix Materials** **Haiqing Lin (Univ. at Buffalo)**	Oral 139 - Strengthening Isoporous Polymer Membranes through Chain Entanglements Adam Mann (UT Austin)	Oral 145 - The application of tangential flow filtration for the continuous processing of monoclonal antibodies: process optimization based on particle morphology Ali Behboudi (Penn State University)
3:50 PM	Oral 130 - Evaluation of Commercial Nanofiltration and Reverse Osmosis Membrane Filtration to Remove PFASs: Effects of Transmembrane Pressures, Water Matrices such as Surfactants, and Ion Valency and Temperatures Guangyu Zhu (NJIT)	**4:00 – 4:30 PM** **Oral 136 - Osmoses: Developing Gas Separation Membranes to Enable Sustainable Energy Solutions** **Holden Lai (OSMOSES)**	Oral 140 - Highly Selective Thin B-Oriented MFI Zeolite Membranes on Scalable Modified Stainless-Steel Supports Jerry Lin (ASU)	Oral 146 - A more sustainable all-membrane process for the purification of monoclonal antibodies Cristiana Boi (Univ. of Bologna)
4:10 PM	Oral 131 – A Concise Analytical Model for the System Performance of an RO Desalination Process Lianfa Song (Texas Tech)	**4:30 – 5:00 PM** **Oral 137 - Deep Flue Gas Decarbonization with Novel Membrane Process Designs** **Richard Baker (MTR)**	Oral 141 - Ion-sieving membranes based on Polymers of Intrinsic Microporosity for aqueous organic redox flow batteries Chunchun Ye (Univ. of Edinburgh)	Oral 147 - IgG-BSA Separation and Purification by Internally Staged Ultrafiltration Lixin Feng (NJIT)
4:30 PM	Oral 132 - Novel artificial water channel membranes with improved molecular separation properties in reverse osmosis Maria Di Vincenzo (KAUST)	**5:00 – 5:30 PM** **Oral 138 - Polymer membranes stability under sour-humid natural gas mixtures and their role towards de-carbonization** **Nitesh Bhuwania (Chevron)**	Oral 142 - Role of salt additives in tuning the structure of β-cyclodextrin-based membranes Jinting Hu (Kuleuven)	Oral 148 - Effect of protein on the location of virus entrapment during virus filtration Ranil Wickramasinghe (Univ. of Arkansas)
4:50 PM	Oral 133 - Regulating interfacial polymerization via multi-functional calcium carbonate based interlayer for highly permselective nanofiltration membrane Mengyang Hu (Kobe University)		Oral 143 - Anti-swelling Characteristics Enhanced via Self-Cross-Linking in MXene-Interlaced Graphene Oxide Membranes Saurabh Tiwary (U of Houston)	Oral 149 - The Effects of Pressure and Process Interruption on The Location of Virus Particles Captured during Virus Filtration Wenbo Xu (Univ. of Arkansas)
5:10 PM	Oral 134 - Three-dimensional modeling of flow and mass transfer in spacer-filled channels in Reverse osmosis/Nanofiltration processes Santiago Cespedes (KAUST)		Oral 144 – Revealing Molecular Mechanisms in Hierarchical Nanoporous Carbon via Nuclear Magnetic Resonance Haiyan Mao (Stanford University)	Oral 150 – Bioseparation Membranes from Crosslinking of Pluronic F127 Self-assembly Seyed Mostafa Tabatabaei (Univ. of Oklahoma)

Wednesday Morning

Parallel Sessions	Room: Milagro-Kearny Membranes for carbon capture II	Room: Peralta-Lamy Specific Ion separations	Room: O'Keeffe Scaling up: bridging the gap between lab and industry	Room: Coronado/DeVargas Contaminants removal from water I
Chairs	David Hopkinson david.hopkinson@netl.doe.gov Joshua D. Moon Joshua.moon@ufl.edu Zhongyun Liu zhongyunliu@gatech.edu	Ngoc Bui ngocbui21@ou.edu Geoffrey Geise geise@virginia.edu	Oguz Karvan o.karvan@emi-twente.nl Guy Ramon ramong@technion.ac.il	Zora Huang zihanhuang@chevron.com Jian Xu jian.xu@bayer.com Reza Foudazi rfoudazi@ou.edu
8:30 AM	Oral 151 – Large pilot testing of MTR's membrane-based post-combustion CO2 capture process Witopo Salim (MTR)	Oral 158 - Elucidating selectivity mechanisms of anionic pollutants in nanofiltration membranes Idit Shefer (Technion)	Oral 165 - Electrosprayed membranes for CO2-selective gas separations Noah Ferguson (UCONN)	Oral 172 - Structure-property relationships in intercalated graphene oxide nanofiltration membranes for enhanced separation of salts and organics from aqueous streams Muskan Sonker (GaTech)
8:50 AM	Oral 152 – Designing membranes using bottlebrush poly(1,3-dioxolane) acetate for CO2 capture Fathy Attia (Univ. at Buffalo)	Oral 159 - A study of NF membrane assessment as pretreatment for lithium recovery from salt-lake brine Juan Zhai (Texas Tech)	Oral 166 - Simplified Synthesis Methods for Membrane-Protein Nanosheets and Biomimetic Membranes Ronald Vogler (UT Austin)	Oral 173 - Removal of Dodecyl Benzenesulfonic Acid by Amine-Modified Membrane Adsorbers Hadi Rouhi (Univ. of Alabama)
9:10 AM	Oral 153 - Random Poly(1,3-dioxolane)-based Terpolymers for CO2 Separations Justin Rosenthal (UT Austin)	Oral 160 - Understanding the dopant effects on the selective transport of cations in conductive polymers Meng Shen (CSU)	Oral 167 - Evaluation of spacer-induced hydrodynamic mixing using particle image velocimetry: Impact on membrane distillation performance Mahdi Malmali (Texas Tech)	Oral 174 - Thin Film Nanocomposite Membranes Embedded with ZIF-67 for Desalination and Boron Removal Suer Kurklu Kocaoglu (UT Austin)
9:30 AM	Oral 154 - Highly Permeable Rubbery Thin Film Composite Membranes for CO2 Capture from Steel Mills Lingxiang Zhu (NETL-DOE)	Oral 161 - Ion-selective Membranes for Critical Materials Separation Ngoc Bui (Univ. of Oklahoma)	Oral 168 - A Fractal Analysis Based Metrology to Quantify Porous Structure of Membranes Riccardo Lovison (UC Boulder)	Oral 175 - Unlocking the power of a gutter layer: A new frontier in dye-desalination using amine-crosslinked nanoporous graphene oxide (aHGO) platform. Ameya Tandel (Univ. at Buffalo)
9:50 AM	Oral 155 - Investigating PIM-1 and cPIM-1 applications for CCUS under real-world conditions Sean R McIntyre (Surface Measurement Systems)	Oral 162 – Aligned macrocycle pores in ultrathin films for accurate molecular sieving Zhiwei Jiang (QMul)	Oral 169 - Design of Experiments to Optimize the Polyimide Membrane Structure Margaret Lee (PALL)	Oral 176 - Experimental and Modeling Results for the Rejection of Ultrashort Per-and Polyfluoroalkyl Substances Aron Griffin (Mines)
10:10 AM	Oral 156 - Retarded O2 transport in Co2+-coordinated supramolecular polymer networks for membrane CO2/O2 separations Haiqing Lin (Univ. at Buffalo)	Oral 163 - Tailoring Interlayer Spacing and Surface Chemistry of MXene Free-standing Membranes for Selective Separation of Precious Ions from Water Ahmed Al Mayyahi (Univ. of Alabama)	Oral 170 - Spiral Wound Module Development and Scale-up for Olefin-Paraffin Separations with OptipermTM Membrane System Jacob Wargo (CoimpactMembrane)	Oral 177 - Polymer evolution and fabrication of PVDF membranes for contaminant removal from water Rachel Kaplan (Univ. of Kentucky)
10:30 AM	Oral 157 – A Membrane Contactor Enabling Energy-Efficient CO2 Capture from Point Sources with Physical Solvent Nicholas Gregorich (ORNL)	Oral 164 - Selective ion conducting polymers for non-aqueous redox flow battery applications Geoffrey Geise (Univ. Of Virginia)	Oral 171 - Effect of polymer molecular weight on the long-term process stability of crosslinked polybenzimidazole (PBI) membranes Adam Oxley (Exactmer)	Oral 178 – Zr-based MOF-NF-Membrane for Selective Removal of Selenium from Produced water Sweta Modak (Univ. of Alabama)

Parallel Sessions	Room: Milagro-Kearny Mixed matrix and composite membranes for gas separation	Room: O'Keeffe Solute selective separations modeling	Room: Peralta-Lamy NAMS Keynote II Talks in this session have a 40 min slot	Room: Coronado/DeVargas Contaminants removal from water II
Chairs	Shailesh Singh Dangwal dangwals@ornl.gov Jay (Junqiang) Liu jliu7@dow.com Hani A.E. Hawa hhawa@precision-combustion.com	Meng Shen meshen@fullerton.edu Jianjia Yu jianjia.yu@nmt.edu	Mahdi Malmali mahdi.malmali@ttu.edu Grigorios Panagakos gpanagak@andrew.cmu.edu Shouliang Yi shouliang.yi@hotmail.com	Zora Huang zihanhuang@chevron.com Yuexiao Shen Yuexiao.Shen@ttu.edu
1:00 PM	Oral 179 - UV-Crosslinked Nanocomposite Membranes for Fast and Selective Transport of Small Molecules Joo-Eon Kim (Krict)	Oral 185 - Ion and water dynamics in the transition from dry to wet conditions in salt-doped PEG Nico Marioni (UT Austin)	**1:00 – 1:30 PM** **Oral 191 - Dirt and waste as source materials for resource recovery membranes** **Seth Darling (ANL)**	Oral 195- Real-Time Induced Magnetic Vibrational Based Antifouling Mechanism for Ultrafiltration (UF) membrane Jasneet Pala (Univ. of Alabama)
1:20 PM	Oral 180 - Computational insights from the polymer/MOF interface in Mixed-Matrix Membranes Moein Hajian Zaker Damghan (Texas Tech)	Oral 186 - The influence of Donnan – Manning and dielectric exclusion on ion partitioning in sulfonated polysulfone Sean Bannon (UVA)	**1:30 – 2:00 PM** **Oral 192 - Pall's Filtration Solution for Separation and Purification in Enabling Decarbonization** **Varun Vakharia (PALL)**	Oral 196 - Catalytic Membranes Based on Reduced Graphene Oxide for Micropollutant Treatment Enabled by Peracetic Acid Erda Deng (Univ. at Buffalo)
1:40 PM	Oral 181 - Leveraging anomalous gas transport in thin-film composite membranes based on polymers of intrinsic microporosity Taehoon Lee (MIT)	Oral 187 - Influence of backbone rigidity and water content on lithium/magnesium selectivity in hydrated polymer membranes Paul Irving (UT Austin)	**2:00 – 2:30 PM** **Oral 193 - Opportunities and challenges for CO2 separation membranes in decarbonization of the electric power, cement, and steel industries** **David Hopkinson (NETL-DOE)**	Oral 197 - Designing the fluorinated superfine-activated carbon adsorptive membrane for selective removal of long- and short-chain PFAS Medha Kasula (Univ. of Alabama)
2:00 PM	Oral 182 - CO2/C2H2 Transport in SSZ-13 Zeolite Crystals and Membranes, and the Implication on Zeolite Membrane Microstructure Shaowei Yang (csuohio)	Oral 188 - Molecular Dynamics Simulations of Lithium/Magnesium Mixed Solutions in Polyamide Membranes Richard Lueptow (Nothwestern Univ.)	**2:30 – 3:00 PM** **Oral 194– Carbon capture through membranes: leveraging multiphysics modeling, dimensional analysis and machine learning to scale up and optimize devices and processes for decarbonization** **Grigorios Panagakos (Carnegie Mellon)**	Oral 198 - Dynamic ammonium retention for nutrient separation from manure digestate Marrit van der Wal (TUE)
2:20 PM	Oral 183 - Gas Permeation and Separation Characteristics of Microporous TpHz COF Membranes Synthesized by Substrate-Assisted Interfacial Polymerization Jose E Lopez-Cazares (ASU)	Oral 189 - Modeling polymer aging in solvent environments with symbolic regression Matthew Rivera (MIT)		Oral 199 - Advanced functionalized polyether ether ketone/polysulfone membranes with 2D nanomaterials for enhanced water treatment performance Usman Yousaf (Univ. of Kentucky)
2:40 PM	Oral 184- Towards the Synthesis of Mixed Matrix Membranes Using Vapor Phase Infiltration Seancarlos Gonzales (Univ. Of Washington)	Oral 190 - Ion-ion Separations in Biomimetic Water Channels Tyler J Duncan (UT Austin)		Oral 200 - Distributed Membrane-Based Water Desalination and Purification Systems: Automation, Data-Driven Operational Models, and Performance Forecasting Nora Marki (University of California, Los Angeles)

Sweeney E-F, 11:00 am - 12:00 pm, Monday, May 13th

Membranes: Smart Enabler for a Sustainable Gas Economy

Plenary Speaker

Dr. Goetz Baumgarten

Head of the Membranes Innovation Growth Field, Evonik

After studying chemistry at the University of Hannover and earning a doctorate in the treatment of landfill leachate with membrane processes, Dr. Baumgarten began his career at Amafilter Deutschland GmbH in Düsseldorf in 1997 as project manager. From 2001, he was product manager for the Membrane Technology Product Line for the entire Amafilter Group (today: Mahle Industrial Filtration) before moving to Evonik in 2005. At Evonik, Dr. Baumgarten initially headed the Membrane Technology Group in the Process Technology & Engineering Service Unit before assuming the position of Business Manager SEPURAN in July 2010. Since January 2016, he has been responsible for the Membrane Business of Evonik.

Abstract

Backward integration is the key driver of Evonik's innovative strength in membranes. Thanks to the group's many years of expertise in polymer chemistry and backward integration, Evonik adjusts the membrane properties already at the development stage of the base material—the high-performance polymer—to produce particularly selective, productive and robust membranes that can withstand extreme pressures and temperatures in challenging conditions.

SEPURAN® membranes are polymer-based hollow-fibre membranes designed to separate gases such as methane (CH_4), nitrogen (N_2), or hydrogen (H_2) from gas mixtures. The advantages of Evonik's membrane technology are more precise separation of the gases, higher productivity and their outstanding robustness and lifetime based on the entirely crosslinked polymer basis. SEPURAN® Green membranes and our 3-stage upgrading process shaped the market for biogas upgrading with now more than 1000 references worldwide. SEPURAN® N_2 membranes for efficient nitrogen generation are used, for example, to inert aircraft tanks. SEPURAN® Noble membranes extract hydrogen transported through natural gas pipelines selectively from the CH_4/H_2 gas mixture at the delivery points. SEPURAN® NG as well as PURAMEM® membranes enable efficient natural gas processing from gas sources with high CO_2 and/or H_2S concentration. SEPURAN® and PURAMEM® membranes are being successfully used in more than 2,000 system installations worldwide.

Following the Sustainable Development Goals of the United Nations, Sustainability is the key growth driver and the cornerstone of Evonik's product portfolio, investments and innovation management and thus an integral part of Evonik's strategy. With the innovative gas separation SEPURAN®, PURAMEM® as well as the advanced ion conducting DURAION® membrane technologies, the specialty chemicals company makes the transition of today's conventional energy supply to a sustainable gas economy of tomorrow possible.

Sweeney E-F, 11:00 am - 12:00 pm, Tuesday, May 14th

Olefin/Paraffin Separations with Supported Ionic Liquid Membranes

Plenary Speaker

Dr. Joan Brennecke

Professor, Cockrell Family Chair in Engineering # 16

Department of Chemical Engineering

The University of Texas at Austin, USA

Dr. Joan Brenneckee is currently Cockrell Family Chair in Engineering #16 in the McKetta Department of Chemical Engineering at the University of Texas at Austin. She began her academic career at the University of Notre Dame after completing her Ph.D. and M.S. (1989 and 1987) degrees at the University of Illinois at Urbana-Champaign and her B. S. at the University of Texas at Austin (1984). Her research interests are primarily in the development of less environmentally harmful solvents. These include supercritical fluids and ionic liquids. In developing these solvents, Dr. Brennecke's primary interests are in the measurement and modeling of thermodynamics, thermophysical properties, phase behavior and separations. Major awards include the 2001 Ipatieff Prize from the American Chemical Society, the 2006 Professional Progress Award from the American Institute of Chemical Engineers, the J. M. Prausnitz Award at the Eleventh International Conference on Properties and Phase Equilibria in Greece in May, 2007, the 2008 Stieglitz Award from the American Chemical Society, the 2009 E. O. Lawrence Award from the U.S. Department of Energy, and the 2014 E. V. Murphree Award in Industrial and Engineering Chemistry from the American Chemical Society. She served as the Editor-in-Chief of the Journal of Chemical & Engineering Data from 2010-2020. Her 250+ research publications have garnered over 24,000 citations (ISI), with an H-index of 66. She was inducted into the National Academy of Engineering in 2012 and served as the Chair of The National Academies study on A Research Agenda for a New Era in Separations Science, published in 2019.

Abstract

Olefins are a key building block of the petrochemicals industry because they are precursor materials for numerous chemical products and plastics. The current commercial production method is steam cracking of ethane followed by cryogenic distillation of olefin/paraffin mixtures, which involves large energy consumption and greenhouse gas footprint (0.3% of world energy use). Membranes using facilitated transport show promising olefin/paraffin selectivity due to the presence of carriers that specifically complex with olefins. We have developed facilitated transport-based supported ionic liquid membranes (SILMs) with excellent olefin/paraffin selectivity and permeability, using mixtures of silver salts and ionic liquids (ILs). The metal ions dissolved in the IL act as carriers and the IL medium allow the carriers to be mobile in the membrane pores. Previously, facilitated transport-based membranes for olefin/paraffin separation have not been viable because the silver carriers deactivate rapidly in the presence of any H2, which chemically reduces the silver salts to inactive silver metal. We show that the proper choice of IL results in stable olefin/paraffin separation performance even after extended exposure to H2. Using Raman spectroscopy, we show how the formation of anion aggregates play a key role in the suppression of silver metal nanoparticle formation. Moreover, we have modified the anodic aluminum oxide (AAO) membrane supportss with uniformly sized ~6-8 nm diameter cylindrical silica mesopores to improve the blow-out transmembrane pressure (TMP). This modification increases the feasible TMP for various ionic liquids from ~4 bar to more than ~20 bar. Additionally, the liquid in the membrane at depths where the AAO membrane has not been modified (i.e., greater than a few microns) is expelled from the large pores, reducing the effective thickness of the liquid in the membrane, and increasing permeance. This approach is analogous to asymmetric polymeric membranes. This innovation paves the way for selective olefin/paraffin separation with dramatically reduced energy consumption, that is viable from laboratory to commercial scale operation.

Sweeney E-F, 11:00 am - 12:00 pm, Wednesday, May 15th

Beyond the upper limit? What we are missing & challenges

Plenary Speaker

Dr. Ho Bum Park

Professor and Department Chair, Energy Engineering
Hanyang University, South Korea

Dr. Park is a Professor in the Energy Engineering Department at Hanyang University (HYU) in Seoul in Korea. He received his Ph.D. in Chemical Engineering from HYU in 2002 and worked as a postdoctoral fellow for the University of Texas at Austin (UT Austin) between 2005 and 2008. He joined HYU as an assistant professor in 2008, an associate professor in 2011, and a full professor in 2017. He worked as a visiting scholar at the Department of Chemical Engineering of UT Austin, USA (2014). He has established a state-of-the-art research laboratory for advanced membrane research at HYU since 2008 and a dozen research collaborations and international scholar exchange programs. During his service period at HYU, he has published more than 140 peer-reviewed SCI papers, including some of the highest-impact scientific journals, contributed 8 book chapters, and more than 90 patents. He has served on the Membrane Society of Korea (MSK) Board of Directors for 14 years. He now serves as a Department Chair of Energy Engineering at Hanyang University. He has organized many international and national conferences and also served as an editorial board member of five scientific journals related to chemical engineering and materials science and engineering.

Abstract

Dr. Robeson first presented the correlation between permeability and selectivity of polymeric materials for various gas pairs of interest in 1991. This brought awareness to the separation limits of polymeric materials for gas separation membranes and motivated much research to overcome them. Subsequently, this concept has also been applied to other types of membranes, such as ultrafiltration, ion exchange, and desalination membranes. These membranes are evaluated based on their water permeance, solute rejection, ion conductivity, water uptake, water permeability, and water/ion selectivity. This has led to a renewed focus on developing membrane materials that can surpass their respective separation limits. Currently, the range of materials that can be used as membranes extends beyond polymers. New materials such as carbons, silica, zeolites, metal-organic frameworks, and 2D materials are being investigated as separation membranes. These materials exhibit superior permeability and selectivity properties that surpass the upper limits. It is important to note that all evaluations are objective and clearly marked as such. Although there are numerous excellent membrane materials available, only around ten are currently used in membrane processes. This presentation will address some of the crucial aspects that are lacking in membrane R&D and the challenging work that lies ahead. Based on my previous experience in R&D for gas separation and other membranes, this presentation discusses the current status of various membrane materials that have been and are being developed. It also comments on the many factors that need to be considered before a material becomes a viable separation membrane in the near future.

Oral Sessions – Monday Morning

Presenting authors are indicated by an asterisk*.

Electrochemical Separations

8:30 AM – 10:50 AM

Room: Coronado/DeVargas

Session Chairs

Caleb Funk
DuPont

Abhishek Narayan Mondal
Donaldson

8:30 AM
Oral 1 - Understanding Polyfluoroalkyl Substance Selectivity in Anion Exchange Membrane Systems

Sam S Thompson (University of Kentucky)*; Angela Gutierrez (University of Kentucky); Rollie Mills (U. Kentucky); Dibakar Bhattacharyya (U. Kentucky)

8:50 AM
Oral 2 - Nanofiltration membranes with thermo- and pH-responsive behavior from lyotropic liquid crystal templating

Younes Saadat (University of Oklahoma); Kyungtae Kim (Los Alamos National Laboratory); Reza Foudazi (University of Oklahoma)*

9:10 AM
Oral 3 - Charge-patterned nanofiltration membranes with immobilized polystyrene sulfonate particles and polyethylene imine

Ali Behboudi (The Pennsylvania State University)*; Philipp Jahn (Lehrstuhl für Technische Chemie II, Universität Duisburg-Essen, 45141 Essen, Germany); Mathias Ulbricht (Lehrstuhl für Technische Chemie II, Universität Duisburg-Essen, 45141 Essen, Germany)

9:30 AM
Oral 4 - Treating Acidic Metal Wastewaters with Ceramic Ion Exchange Membranes

Emily Rabe (Membrion)*; Rachel Malone (Membrion); Danny Bondurant (Membrion); Stephanie Candelaria (Membrion); Olivia Lenz (Membrion); Gregory M Newbloom (Membrion, Inc.)

9:50 AM
Oral 5 - Synthesis and Testing of Electrochemical Activity in a Polyether Sulfone-Based Ultrafiltration Membrane for Contaminant Removal

Utkarsh Misra (IIT Bombay)*; Kritika Jashrapuria (Indian Institute of Technology Bombay); Swatantra Pratap Singh (IIT BOMBAY)

10:10 AM
Oral 6 - Electroconductive Filters and Membranes for Desalination and Wastewater Treatment

Swatantra Pratap Singh (IIT BOMBAY)*

10:30 AM
Oral 7 - Performance of Commercial Bipolar Membranes for Ocean Alkalinity Enhancement Application

Bassel A. Abdelkader (McMaster University)*; Charles-François de Lannoy (McMaster University)

Advanced Membranes for Gas and Vapor Separation I

8:30 AM – 10:50 AM

Room: Milagro-Kearny

Session Chairs

Nitesh Bhuwania
Chevron

Chen Zhang
University of Maryland

Simona Liguori
Clarkson University

8:30 AM
Oral 8 - Enhanced H2 transport in polybenzimidazole by blending with a highly permeable polyimide

Narjes Esmaeili (The State University of New York at Buffalo)*; Leiqing Hu (University at Buffalo); Erda Deng (State University of New York at Buffalo); Haiqing Lin (The State University of New York at Buffalo)

8:50 AM
Oral 9 - Tightening ultramicropores of structurally engineered carbon molecular sieve membranes for precise gas separation
Zhongyun Liu (Georgia Institute of technology)*; William Koros (Georgia Institute of Technology)

9:10 AM
Oral 10 - Computer vision-based measurement of pure- and mixed-gas swelling in polymer membranes exhibiting configurational free volume
William Box* (University of Oklahoma); Lucas Condes (University of Oklahoma); Matthew T Webb (University of Oklahoma); Agboola OS Suleiman (University of Notre Dame); Michele Galizia (University of Oklahoma); Ruilan Guo (University of Notre Dame)

9:30 AM
Oral 11 - Hydrogen- and acetylene-stable silver ion facilitated transport membranes: the effect of salt content on fractional free volume and gas permeability
Matthew N Davenport (The University of Texas at Austin)*; Tangqiumei Song (UT Austin); Justin J Rosenthal (The University of Texas at Austin); Cara Doherty (CSIRO); Joan Brennecke (The University of Texas at Austin); Benny Freeman (The University of Texas at Austin)

9:50 AM
Oral 12 - Exploring the separation efficiency of ultra-selective 3D CANAL hydrocarbon polymeric membranes for hydrogen-based separations after long-term aging
Jing Ying Yeo (Massachusetts Institute of Technology)*; Francesco Maria Benedetti (Massachusetts Institute of Technology); Zachary Smith (Massachusetts Institute of Technology)

10:10 AM
Oral 13 - Evaluation of air addition into methane dry reforming reaction in Pd/YSZ membrane reactor for syngas production
Omid Jazani (Clarkson University)*; Simona Liguori (Clarkson University)

10:30 AM
Oral 14 - Synthesis and Characterization of ZnCl2-Immobilized Molten Salt (IMS) membrane for NH3 separation
Moses T Adejumo (Clarkson University)*; Simona Liguori (Clarkson University)

Functional Membrane Fabrication and Environmental Sustainability

8:30 AM – 10:50 AM

Room: O'Keeffe

Session Chairs
Margaret Lee
Pall Corporation

Priyesh Wagh
The Lubrizol Corporation

8:30 AM
Oral 15 - Scaled Manufacturing of Supported Loose-NF Membranes with Eco-Friendly Solvents for Water Treatment Applications
David Lu (University of Kentucky)*; Kwangjun Jung (Georgia Institute of Technology); Ju Young Shim (Georgia Institute of Technology); Tequila Harris (Georgia Institute of Technology); Isabel Escobar (University of Kentucky)

8:50 AM
Oral 16 - Sustainable additive manufacturing of polysulfone membranes for liquid separations
Brian Leonard (West Virginia University)*; Oishi Sanyal (West Virginia University); Harrison Loh (West Virginia University); Kostas Sierros (West Virginia University); David Lu (University of Kentucky); Ebuka Ogbuoji (University of Kentucky); Isabel Escobar (University of Kentucky)

9:10 AM
Oral 17 - An alternative application of 2D carbon-based membranes for efficient air-separation and H2 purification from gas mixtures
Dániel Gardenö (UCT Prague)*; Karel Friess (UCT Prague); Zdenek Sofer (UCT Prague); Malte Kaspereit (FAU Erlangen-Nürnberg)

9:30 AM

Oral 18 - Understanding the mechanism of additive-assisted isoporous membrane formation through X-ray scattering

Noah P Wamble (University of Texas at Austin)*; Mostafa Nassr (UT Austin); Matthew Landsman (Lawrence Berkeley National Lab); Louise Kuehster (University of Texas at Austin); Joan Brennecke (The University of Texas at Austin); Nathaniel Lynd (The University of Texas at Austin); Gregory Su (Lawrence Berkeley National Lab); Benny Freeman (The University of Texas at Austin)

9:50 AM

Oral 19 - Elucidating the Structure-Property-Performance Relationship of Incorporating Waste Polymethyl Methacrylate in Polyvinylidene Fluoride Membrane Fabrication

Joseph T Danner (The University of Alabama)*; Liam Papier (The University of Alabama); Steven T Weinman (The University of Alabama)

10:10 AM

Oral 20 - Enhanced bipolar membrane water disassociation reaction with layer-by-layer assembled clay composite junctions

Nadia Boulif (Eindhoven University of Technology)*; Menno Houben (Eindhoven University of Technology); Zandrie Borneman (Eindhoven University of Technology); Kitty Nijmeijer (Eijndhoven)

10:30 AM

Oral 21 - Advancing sustainable membrane fabrication by investigating the effect of biodegradable and non-biodegradable solvents on polylactic acid (PLA) membranes

Escobar, Isabel (University of Kentucky)*; Ogbuoji, Ebuka (University of Kentucky)

Process Intensification and Decarbonization

8:30 AM – 10:50 AM

Room: Peralta-Lamy

Session Chairs

Albert Kim
University of Hawai'i

Syed Islam
Oak Ridge National Laboratory

8:30 AM

Oral 22 - Zeolite Membrane Reactor for CO2 Hydrogenation to Methanol: A Simulation Study on Comparison of Water and Methanol Selective Membranes

Sieun Jung (Korea University/Arizona State University); Mahsa Sharafi (Arizona State University); Jerry Lin (Arizona State University)*

8:50 AM

Oral 23 - Catalytic Membrane Reactors for combined separation and contaminant degradation

Nan Zhang (Tianjin University); Yichen Wu (McMaster University); Gavin Yuen (McMaster University); Abdelrahman Awad (McMaster University); Emmanuel Doelsch (Universite Montpellier); Charles-François de Lannoy (McMaster)*

9:10 AM

Oral 24 - Computational Fluid Dynamics and Systems Engineering to minimize membrane-based carbon capture costs

Hector A Pedrozo (Carnegie Mellon University)*; Cheick Dosso (Carnegie Mellon University); Thien Tran (National Energy Technology Laboratory); Lingxiang Zhu (National Energy Technology Laboratory); Victor A Kusuma (NETL); David Hopkinson (NETL); Lorenz Biegler (Carnegie Mellon University); Grigorios Panagakos (Carnegie Mellon University)

9:30 AM

Oral 25 - Innovating Membranes for Carbon Capture: From Material Synthesis to Process Prototype

Yang Han (The Ohio State University)*; Yutong Yang (The Ohio State University); Jingying Hu (National Renewable Energy Laboratory); Ruizhi Pang (The Ohio State University); Babul Prasad (The Ohio State University); Winston Ho (OSU)

9:50 AM

Oral 26 - Advancements of Responsive and Catalytic Membrane Reactor for Water and Halo-Organics Remediation

Dibakar Bhattacharyya (U. Kentucky)*; Rollie Mills (U. Kentucky); Francisco Leniz (U. Kentucky); Sam S Thompson (University of Kentucky); Subrajit Bosu (U. Kentucky)

10:10 AM

Oral 27 - Intensified propane dehydrogenation in carbon molecular sieve hollow fiber membrane reactors

Lu Liu (University of Maryland); Antara Bhowmick (University of Delaware); Dongxia Liu (University of Delaware); Chen Zhang (University of Maryland)*

10:30 AM

Oral 28 - Mathematical Modeling & Optimization of a Water-Gas Shift Membrane Reactor for Low-Cost H2 Production with CO2 Capture

Hani A.E. Hawa (Precision Combustion, Inc.)*; Kyle Hawley (Precision Combustion, Inc.); Damian T Agi (University of Notre Dame); Alexander Dowling (University of Notre Dame)

Oral Sessions - Monday Afternoon

Presenting authors are indicated by an asterisk*.

Water Treatment and Desalination Processes

1:45 PM – 3:45 PM

Room: Coronado/DeVargas

Session Chairs

Ngai Yin Yip
Columbia University

Yian Chen
National Renewable Energy Laboratory

1:45 PM
Oral 29 - Revisiting the value of high-flux seawater reverse osmosis membranes with variable-flow energy recovery
Sangsuk Lee (UCLA)*; Eric Hoek (UCLA)

2:05 PM
Oral 30 - Performance and Environmental Evaluation of Biochar-based Membranes Synthesized from Various Petroleum- and Bio-derived Solvents
Abelline K Fionah (University of Kentucky)*; Isaac Oluk (University of Kentucky); Isabel Escobar (University of Kentucky)

2:25 PM
Oral 31 - High Recovery Desalination of High Mineral Scaling Source Water via Integrated RO-NF Concentrator, Concentrate Desupersaturation and Second Stage RO-NF Desalting
Yakubu Jarma (University of California, Los Angeles); Jose Morales (University of California, Los Angeles); Phoebe Strauss (University of California, Los Angeles); Yoram Cohen (Professor)*

2:45 PM
Oral 32 - Electrifying membrane distillations: Localized Joule heating of laser-induced graphene membranes
Md Hasib Al Mahbub (Texas Tech University)*; Mahdi Malmali (Texas Tech University)

3:05 PM
Oral 33 - The analytic solution of interfacial concentration with observed rejection ratio during dead-end membrane filtration
Albert Kim (U. of Hawai'i)*

3:25 PM
Oral 34 - High-Throughput Experimentation for High-Pressure Reverse-Osmosis Membranes
Hongchen Wang (University of Toronto)*

Advanced Materials for Gas and Vapor Separation II

1:45 PM – 3:45 PM

Room: Milagro-Kearny

Session Chairs

Nitesh Bhuwania
Chevron

Chen Zhang
University of Maryland

Simona Liguori
Clarkson University

1:45 PM
Oral 35 - What Membranes Do We Need for Industrial Propylene/Propane Separations?
Yang Liu (MTR)*; Tim Merkel (Membrane Technology & Research Inc.); Richard Baker (Membrane Technology & Research Inc.)

2:05 PM
Oral 36 - Scale-up of perfluoro(butenyl vinyl ether) (PBVE) and perfluoro(2,2-dimethyl-1,3-dioxole) (PDD) copolymers for the separation of R-410A
Abby N Harders (University of Kansas)*; Luke Wallisch (University of Kansas); Michael Lundin (University of Kansas); Chloe Le (California State Polytechnic); Gabby Zaher (University of Washington); Whitney White (Chromis Technologies); Mark Shiflett (University of Kansas)

2:25 PM
Oral 37 - Selective photonic gasification of strained oxygen clusters on graphene for overcoming the trade-off between pore size and density

Luc S Bondaz (Ecole polytechnique fédérale de Lausanne)*; Anshaj Ronghe (Indian Institute of Science); Shaoxian Li (Ecole polytechnique fédérale de Lausanne); Kristians Cernevics (Ecole polytechnique fédérale de Lausanne); Jian Hao (Ecole polytechnique fédérale de Lausanne); Oleg Yazyev (Ecole polytechnique fédérale de Lausanne); Ganapathy Ayappa (Indian Institute of Science); Kumar Varoon Agrawal (Ecole polytechnique fédérale de Lausanne)

2:45 PM
Oral 38 - Pall Filtration is Enabling the Energy Transition

Soondeuk Jeung (Pall Corporation)*; Kyle Kathan (Pall Corporation); Varun Vakharia (Pall Corporation); Maria Lingerfelt (Pall Corporation)

3:05 PM
Oral 39 – New class of imidazolium-based ionenes for gas separation membranes

Sudhir Ravula (University of Alabama)*; Jason Bara (University of Alabama)

3:25 PM
Oral 40 - Adjustable microstructure design of a polyphthalonitrile precursor and the derived novel carbon molecular sieve membrane for gas separation

Chun-Po Hu (Nanyang Technological University)*

Hollow Fiber Membranes

1:45 PM – 3:45 PM

Room: O'Keeffe

Session Chairs

Stephen DeWitt
Arkema

Jianjia Yu
New Mexico Tech

1:45 PM
Oral 41 - Experimental and theoretical evaluation of feed flow collar design for shell fed hollow fiber membrane modules

James Tran (UToledo)*; Thien Tran (National Energy Technology Laboratory); Maya Schuchert (Columbia University); Victor A Kusuma (NETL); Lili Sun (University of Toledo); Comfort Oluleke (University of Toledo); Nathan Diemler (NETL); Lingxiang Zhu (National Energy Technology Laboratory); Grigorios Panagakos (Carnegie Mellon University); Glenn Lipscomb (University of Toledo, Toledo); David Hopkinson (NETL)

2:05 PM
Oral 42 - Development of a modular device for evaluating hollow fiber membrane permeance

Timothy Warner (McMaster University)*; Charles De Lannoy (McMaster University); Yichen Wu (McMaster University)

2:25 PM
Oral 43 - Physical Aging of Defect-Free Asymmetric Functionalized PIM Hollow Fiber Membranes

Philippe Jean-Baptiste (Massachusetts Institute of Technology)*; Zachary Smith (MIT)

2:45 PM
Oral 44 - Improved nanofiltration through inverted membrane structures

Tjerk R Watt (University of Twente)*; Wiebe M. de Vos (University of Twente); Joris de Grooth (Univeristy of Twente); Esra te Brinke (University of Twente)

3:05 PM

Oral 45 - Tannic acid-coated hollow fiber for dehydration: From material to membrane and module

Lakshmeesha Upadhyaya (King Abdullah University of Science and Technology (KAUST))*

3:25 PM

Oral 46 - Developing a Thin Film Composite Facilitated Transport Hollow Fiber Membrane for Post Combustion CO2 capture

Oguz Karvan (EMI Twente BV)*; Ilse Pierik (EMI Twente BV); Jurjen Regenspurg (EMI Twente BV); Patrick de Wit (EMI Twente BV); Claudia Ruiz Ruiz (Aqualung Carbon Capture AS); Arun Kumar Panneer Selvam (Aqualung Carbon Capture AS); Paul A. Buchinger (Aqualung Carbon Capture AS); Michael Duthie (Aqualung Carbon Capture UK Limited); Saravanan Janakiram (Aqualung Carbon Capture AS)

Liquid Extraction and Contaminant Separation

1:45 PM – 3:45 PM

Room: Peralta-Lamy

Session Chairs

Yoram Cohen
University of California, Los Angeles

Isabel Escobar
University of Kentucky

1:45 PM

Oral 47 - XpressKleen 1 nm expanded PTFE membrane for harsh chemical filtration

Zhefei Li (Pall Corporation)*; SIAVASH DARVISHMANESH (Pall corporation); Catherine Agcaoili (Pall Corporation)

2:05 PM

Oral 48 - 2,3-Butanediol Separations using Membrane Assisted Liquid-Liquid Extraction

Shailesh Singh Dangwal (Oak Ridge National Lab)*; Syed Islam (Oak Ridge National Laboratory); Blake Trusty (Oak Ridge National Laboratory); Zachary Coin (Oak Ridge National Lab); Ramesh Bhave (Ornl)

2:25 PM

Oral 49 - Continuous membrane demulsification for in situ solvent extraction of carboxylic acid

Yian Chen (National Renewable Energy Laboratory)*; Eric Tan (National Renewable Energy Laboratory); Jacob Dempsey (National Renewable Energy Laboratory); Patrick Saboe (National Renewable Energy Laboratory); Violeta Sànchez i Nogué (National Renewable Energy Laboratory); Jeffrey Linger (National Renewable Energy Laboratory); Gregg Beckham (National Renewable Energy Laboratory)

2:45 PM

Oral 50 - Membrane nanoprecipitation enables the continuous production of uniform nanoparticles at higher concentrations of the polymer solution than with a batch system

Lidietta GIORNO (National Research Council of Italy - Institute on Membrane Technology (CNR-ITM))*; Antonio Condello (National Research Council of Italy - Institute on Membrane Technology (CNR-ITM)); Emma Piacentini (National Research Council of Italy, Insitute on membrane Technology, CNR-ITM)

3:05 PM

Oral 51 - High temperature ceramic membrane separations for carbon circularity: From lube oil purification to chemical recycling of plastics

Pieter Vandezande (VITO)*; Anita Buekenhoudt (VITO); Efecan Pakaner (VITO); Fady Nahra (VITO); Pedro Borges (VITO); Mohammed Nazeer Khan (VITO); Mohammed El Ibrahimi (VITO); Miet Van Dael (VITO)

3:25 PM

Oral 52 - Targeted Separation Scheme of Polyurethane Depolymerization Products

Taysha B Telenar (Arizona State University)*; Timothy Long (Arizona State University); Matthew Green (Arizona State University)

Oral Sessions - Monday Evening

Presenting authors are indicated by an asterisk*.

Desalination and Osmotic Processes I

4:00 PM – 6:00 PM

Room: Coronado/DeVargas

Session Chairs

Jovan Kamcev
University of Michigan

Jonathan Maisonneuve
Oakland Univeristy

Muskan Sonker
Georgia Tech

4:00 PM

Oral 53 - Compaction of high-pressure water treatment membranes: real-time quantification and support layer contribution

Weijian Ding (University of Southern California)*; Kexin Ma (University of Southern California); Amy Childress (University of Southern California)

4:20 PM

Oral 54 - Membrane heterogeneity alters our interpretation of effective energy barriers to transport

Nathanael S Schwindt (University of Colorado Boulder)*; Mor Avidar (Technion); Razi Epsztein (Technion - Israel Institute of Technology); Anthony Straub (University of Colorado Boulder); Michael Shirts (University of Colorado Boulder)

4:40 PM

Oral 55 - Evidence for solution-diffusion in pressure-induced transport of solvent through dense polymer membranes

Kevin Reimund (University of Texas at Austin)*; Rahul Sujanani (The University of Texas at Austin); Jorge Hernandez (The University of Texas at Austin); Kristofer Gleason (The University of Texas at Austin); Manish Kumar (University of Texas at Austin); Benny Freeman (The University of Texas at Austin)

5:00 PM

Oral 56 - Is the Role of Concentration Polarization Overstressed in Membrane Filtration Processes?

Lianfa Song (Texas Tech University)*

5:20 PM

Oral 57 - Modeling and experimental analysis of pH influence in boron separation in reverse osmosis

Fernan David Martinez Jimenez (KAUST)*

5:40 PM

Oral 58 - Flexible RO-NF System Configuration Operation in the Mode of Partial Concentrate Recycle for Inland Water Desalination

Yakubu Jarma (University of California, Los Angeles)*; Beau Akhavain (University of California, Los Angeles); Yoram Cohen (Professor)

Novel Synthetic Strategies for Gas Separations

4:00 PM – 6:00 PM

Room: Milagro-Kearny

Session Chairs

Victor Kusuma
National Energy Technology Laboratory

Ali Rownaghi
National Energy Technology Laboratory

Jaason Williams
Shell

4:00 PM

Oral 59 - Permeance Prize: Utilizing Molecular Design to Integrate Ionic Liquids within High Performance Polymer Gas Separation Membranes: Unprecedented Opportunities for Control over Material Properties and Performance

Jason Bara (University of Alabama)*

4:20 PM
Oral 60 - Few-Cycle Atomic Layer Deposition to Nanoengineer Polymeric Materials for Efficient Gas Separations

Leiqing Hu (University at Buffalo)*; Won Il Lee (Stony Brook University); Ashwanth Subramanian (Stony Brook University); Erda Deng (State University of New York at Buffalo); Kim Kisslinger (Brookhaven National Laboratory); Shouhong Fan (University of Colorado); Vinh Bui (University at Buffalo); Yifu Ding (Membrane Science, Engineering, and Technology (MAST) Center, Department of Mechanical Engineering, University of Colorado Boulder); Chang-Yong Nam (Stony Brook University); Haiqing Lin (The State University of New York at Buffalo)

4:40 PM
Oral 61 - Amine Modified PIM-1 for Highly Selective Olefin/Paraffin Separation

Bo Wei Cynthia Chen (University of Notre Dame)*; Casey O'Brien (University of Notre Dame)

5:00 PM
Oral 62 - A New Microcapillary Film Carbon Molecular Sieve Membrane

Jay (Junqiang) Liu (The Dow Chemical Company)*

5:20 PM
Oral 63 - The Effect of Crosslinker Concentration on Competitive Release Kinetics and Physical Characteristics of Thermo-Responsive, Lignin-Based Soft Composites

Missoury Wolff (Clemson University)*; Eric Davis (Clemson University)

5:40 PM
Oral 64 -

Tailored Membrane Formation

4:00 PM – 6:00 PM

Room: O'Keeffe

Session Chairs

Lakshmeesha Upadhyaya
King Abdullah University of Science and Technology

Luis Francisco Villalobos
University of Southern California

Varun Vakharia
Pall Corporation

4:00 PM
Oral 65 - Sulfonated Polyamide Thin Film Composite Membranes for Electrodialysis

Deniz Ipekci (University of Connecticut)*; Yasmin Teixeira Bimbatti (UCONN); Xingyu Wang (University of Connecticut); Inara Oliveira (University of Virginia); Anna Harris (University of Virginia); Geoffrey M Geise (University of Virginia); Baikun Li (University of Connecticut); Jeffrey R McCutcheon (University of Connecticut)

4:20 PM
Oral 66 - Compaction-resistant Thin Film Crosslinked Composite (TFX) Ultra-high Pressure Reverse Osmosis

Jishan Wu (UCLA)*; Eric Hoek (UCLA)

4:40 PM
Oral 67 - CFD modeling of high-flux plate-and-frame membrane modules for industrial carbon capture

Cheick Dosso (Carnegie Mellon University)*; Hector A Pedrozo (Carnegie Mellon University); Thien Tran (National Energy Technology Laboratory); Lingxiang Zhu (National Energy Technology Laboratory); Victor A Kusuma (NETL); David Hopkinson (NETL); Lorenz Biegler (Carnegie Mellon University); Grigorios Panagakos (Carnegie Mellon University)

5:00 PM
Oral 68 - Free Energy Surface Landscape of Interfacial Polymerization for Polyamide Membrane Formation

Cheick Dosso (Carnegie Mellon University)*; Hector A Pedrozo (Carnegie Mellon University); Thien Tran (National Energy Technology Laboratory); Lingxiang Zhu (National Energy Technology Laboratory); Victor A Kusuma (NETL); David Hopkinson (NETL); Lorenz Biegler (Carnegie Mellon University); Grigorios Panagakos (Carnegie Mellon University)

5:20 PM
Oral 69 - The influence of additives on polystyrene membranes created by temperature-induced phase separation

Sulaiman Dhameri (University of Kentucky)*; Ibrahim Imam (University of Kentucky); Usman Abbas (University of Kentucky); Qing Shao (University of Kentucky); Gosia Chwatko (University of Kentucky)

5:40 PM
Oral 70 - The dynamics of Interfacial Polymerization: insights on film stability during formation

Guy Z Ramon (Technion - Israel Institute of Technology)*; Adi Ben Zvi (Technion)

Lithium Separation

4:00 PM – 6:00 PM

Room: Peralta-Lamy

Session Chairs
Ngoc Bui
University of Oklahoma

Harekrushna Behera
University of Texas at Austin

4:00 PM
Oral 71 - Co-Ion Specific Effect on Li Salt Sorption in Cation Exchange Membrane

Rachel Z Huang (UT Austin)*; Lynn Katz (The University of Texas at Austin); Benny Freeman (The University of Texas at Austin)

4:20 PM
Oral 72 - Mixed counter-ion sorption in ion-exchange membranes: understanding the role of membrane water content

Harsh Patel (University of Michigan)*; José Carlos Díaz (University of Michigan); Jovan Kamcev (University of Michigan)

4:40 PM
Oral 73 - Preparation of lithium-ion selective mixed matrix membrane for lithium recovery from real brine solution via diffusion dialysis process

Sandip Pal (University of California Los Angeles)*; David Jassby (UCLA); Eric Hoek (UCLA)

5:00 PM
Oral 74 - Mixed binary alkali halide salt transport in PEGDA

Aubrey Quigley (The University of Texas at Austin)*; Everett S Zofchak (The University of Texas at Austin); Kate Dycha (The University of Texas at Austin); Nathaniel Lynd (The University of Texas at Austin); Benny Freeman (The University of Texas at Austin)

5:20 PM
Oral 75 - Tuning ionic selectivity in nanoporous graphene membranes using atomic layer deposition and polyelectrolyte assembly

Simar K Mattewal (MIT)*; Rohit Karnik (MIT)

5:40 PM
Oral 76 - Lithium Selective Artificial Ion Channels for Lithium Separation

Harekrushna Behera (The University of Texas at Austin)*; Tyler J Duncan (The University of Texas at Austin); Laxmicharan Samineni (University of Texas at Austin); Hyeonji Oh (University of Texas at Austin); Venkat Ganesan (The University of Texas at Austin); Manish Kumar (University of Texas at Austin)

Oral Sessions - Tuesday Morning

Presenting authors are indicated by an asterisk*.

Membranes for Carbon Capture I

8:30 AM – 10:50 AM

Room: Milagro-Kearny

Session Chairs

David Hopkinson
National Energy Technology Laboratory

Joshua D. Moon
University of Florida

Zhongyun Liu
Georgia Tech

8:30 AM
Oral 77 - Leveraging competitive sorption in microporous polymer membranes to enhance CO2- and H2S-specific gas separation performance

Pablo A Dean (Massachusetts Institute of Technology)*; Katherine Mizrahi Rodriguez (MIT); Yifan Wu (Massachusetts Institute of Technology); Zachary Smith (MIT)

8:50 AM
Oral 78 - Amine Functionalized Supported Ionic Liquid Membranes (SILMs) for CO2/N2 Separation

Antoine Chamoun-Farah (University of Texas at Austin)*; Austin Keller (University of Texas at Austin); Mariam Balogun (University of Texas at Austin); Louise Canada (University of Texas at Austin); Joan Brennecke (The University of Texas at Austin); Benny Freeman (The University of Texas at Austin)

9:10 AM
Oral 79 - Effect of Viscosity of Deep Eutectic Solvent on CO2 Capture Performance in an Energy Efficient Membrane Contactor Based Process

Syed Islam (Oak Ridge National Laboratory)*; Zachary Coin (Oak Ridge National Laboratory); Shailesh Singh Dangwal (Oak Ridge National Lab); Mary Irwin (University of Tennessee at Knoxville); Thomas Knight (Vanderbilt University); Ramesh Bhave (Ornl); Gernot Rother (Oak Ridge National Laboratory); Jacek Jakowski (Oak Ridge National Laboratory); Arif Arifuzzaman (Oak Ridge National Laboratory); Vera Bocharova (Oak Ridge National Laboratory); Robert Sacci (Oak Ridge National Laboratory); Ilia Ivanov (Oak Ridge National Laboratory); Jingsong Huang (Oak Ridge National Laboratory); David Sholl (Oak Ridge National Laboratory)

9:30 AM
Oral 80 - Incorporating phenyl groups in polyorganosilica membranes to improve hydrothermal Stability for H2/CO2 Separation

Vinh Bui (University at Buffalo)*; Varun Reddy Satti (University at Buffalo); Elizabeth Haddad (University at Buffalo); Ameya Tandel (State University of New York, Buffalo, NY); Narjes Esmaeili (The State University of New York at Buffalo); Sai Srikar Chundury (University at Buffalo); Lingxiang Zhu (National Energy Technology Laboratory); Haiqing Lin (The State University of New York at Buffalo)

9:50 AM
Oral 81 - Effect of Film Thickness and Temperature on the CO2 Sorption in Poly(ethyleneimine)

John Hoffman (NIST); Avery Baumann (NIST); Christopher M Stafford (NIST)*

10:10 AM
Oral 82 - 100% Carbon Capture from a Carbon Dioxide-Argon Flue Gas using Polaris Membrane

Pingjiao Hao (Membrane Technology & Research Inc.)*; Tim Merkel (Membrane Technology & Research Inc.)

10:30 AM
Oral 83 - Thin Film Composite Membranes from Polymers of Intrinsic Microporosity for CO2 Separation
Sunshine Iguodala (Imperial College London)*

Membrane Fouling

8:30 AM – 10:50 AM

Room: Coronado/DeVargas

Session Chairs
Nima Shahkaramipour
Pentair

Manish Kumar
University of Texas at Austin

8:30 AM
Oral 84 - Optimizing electrospun composite-fiber patterns on nanofiltration membranes for improved biofouling resistance
Amos Taiswa (Montana Tech)*; Jessica Andriolo (Montana Tech); Jack Skinner (Montana Tech)

8:50 AM
Oral 85 - Effect of Pattern Size Relative to Oil Droplet Size on Oil Fouling in Nanofiltration
Lauren M Ward (The University of Alabama)*; Catherine Martin (The University of Alabama); Steven T Weinman (The University of Alabama)

9:10 AM
Oral 86 - The effect of alternating electric fields on membrane colloidal fouling
Elina Yachnin (Technion - Israel Institute of Technology)*; David Jassby (University of California); Tamar Segal-Peretz (Technion-Israel Institute of Technology); Guy Z Ramon (Technion - Israel Institute of Technology)

9:30 AM
Oral 87 - Optimization of Fouling Resistant Polyelectrolyte Coacervate Membranes via Aqueous Phase Separation
Shao-Hsiang Hung (University of Massachusetts Amherst)*; Jessica Schiffman (University of Massachusetts Amherst)

9:50 AM
Oral 88 - Mechanism of crystallization and membrane scaling mitigation via carbon nanotube-containing spacer in membrane process
Seongeom Jeong (Pusan National University)*; Sanghyun Jeong (Pusan National University)

10:10 AM
Oral 89 - Atomic layer processes for membrane fabrication and surface modification
Brian Welch (Technion- Israel Institute of Technology); Ruoke Cai (Technion- Israel Institute of Technology); Assaf Simon (Technion); Zhenzhen Zhang (Helmholtz-Zentrum Hereon); Volker Abetz (Helmholtz-Zentrum Hereon); Tamar Segal-Peretz (Technion- Israel Institute of Technology)*; Clarissa Abetz (Helmholtz-Zentrum Hereon)

10:30 AM
Oral 90 - Fabrication of polymer brush grafted anti-biofouling ultra-filtration membrane using atom transfer radical polymerization
Kritika Jashrapuria (Indian Institute of Technology Bombay); Swatantra Pratap Singh (IIT BOMBAY)*

Organic Molecules and Solute Selective Separations

8:30 AM – 10:50 AM

Room: Peralta-Lamy

Session Chairs
Geoff Geise
University of Virginia

Ryan Lively
Georgia Tech

8:30 AM
Oral 91 - Organic Solvent Reverse Osmosis Separations of Hydrocarbon-Alcohol Mixtures using Matrimid Membranes
Bennett D Marshall (ExxonMobil)*; Ryan Lively (Georgia Tech); Youngjoo Lee (Georgia Tech); JR Johnson (Exxon Mobil); David Moser (ExxonMobil)

8:50 AM
Oral 92 - Fluorine-Rich Poly(Arylene Amine) Membranes for the Separation of Sustainable Fuels and Hydrocarbons
Yi Ren (Georgia Institute of Technology)*; Ryan Lively (Georgia Tech); Sheng Guo (Nanjing University); Hui Ma (Nanjing University)

9:10 AM
Oral 93 - Organic solvent nanofiltration and reverse osmosis performance of carboxylated PIM-1 in neat and inter-crosslinked mixed membrane configurations
Lucas Condes (University of Oklahoma)*; Matthew T Webb (University of Oklahoma); Will Box (University of Oklahoma); Michele Galizia (University of Oklahoma)

9:30 AM
Oral 94 - Volatile fatty acid selectivity during electrodialysis: role of solution composition on volatile fatty acid partitioning and mobility in an anion exchange membrane
Holly M Haflich (University of North Carolina-Chapel Hill)*; Joshua Singleton (University of North Carolina at Chapel Hill); Orlando Coronell (University of North Carolina at Chapel Hill)

9:50 AM
Oral 95 - Membrane-based Ultrapurification of Isopropyl Alcohol
Kamalesh K Sirkar (New Jersey Institute of Technology)*; John Chau (New Jersey Institute of Technology)

10:10 AM
Oral 96 - Co-Transport of Alcohols and Carboxylate Ions in Sulfonated Ion Exchange Membranes with PEGPEA as a Blocking Group
Pravin Parasakthi Aravindhan (Auburn University)*; Bryan S Beckingham (Auburn University); Jung Min Kim (University of Delaware)

10:30 AM
Oral 97 - Fundamental investigation about the effect of chemical structure of foulant on fouling behavior of polyamide membrane in organic solution
Atsushi Matsuoka (Kobe University)*; Ryouhei Aso (Kobe University); Eiji Kamio (Kobe University); Yasunao Okamoto (Kobe University); Tomohisa Yoshioka (Kobe University); Keizo Nakagawa (Kobe University); Hideto Matsuyama (Kobe University)

Membrane Distillation and Pervaporation

8:30 AM – 10:50 AM

Room: O'Keeffe

Session Chairs

Amos Taiswa
Montana Technological University

Allyson McGaughey
The University of New Mexico

Lucy-Mar Camacho
Texas A&M University Kingsville

8:30 AM

Oral 98 - Long-term Operation of Pilot-scale Membrane Bioreactor – Membrane Distillation for Potable Water Reuse

Luke Presson (University of Arizona)*; Michael Hegetschweiler (University of Arizona); Andrea Achilli (University of Arizona)

8:50 AM

Oral 99 - Combined chlorine dioxide–membrane distillation for the treatment of produced water

Mahdi Malmali (Texas Tech University)*; I-Min Hsieh (Texas Tech University)

9:10 AM

Oral 100 - Membrane Distillation and Crystallization for Treating Hydraulic Fracturing Produced Water

Wickramasinghe, Ranil (University of Arkansas)*

9:30 AM

Oral 101 - Advances in the Characterization and Digital Reconstruction of 2-D and 3-D Pore Structure

Jay Werber (University of Toronto)*; Sima Zeinali (University of Toronto); Hooman Chamani (University of Toronto); Arash Rabbani (University of Leeds); Patrick Lee (University of Toronto); Jason R Hattrick-Simpers (University of Toronto)

9:50 AM

Oral 102 - Fast and selective desalination using pressure-driven distillation with nanoscale air gaps

Anthony Straub (University of Colorado Boulder)*

10:10 AM

Oral 103 - From surface wetting to internal wetting: wetting front progression in membrane distillation

Shounak G Joshi (University of Southern California)*; Amy Childress (University of Southern California); Allyson McGaughey (University of New Mexico)

10:30 AM

Oral 104 -

Oral Sessions - Tuesday Afternoon

Presenting authors are indicated by an asterisk*.

Desalination and Osmotic Processes II

1:00 PM – 3:00 PM

Room: Coronado/DeVargas

Session Chairs

Jovan Kamcev
University of Michigan

Jonathan Maisonneuve
Oakland University

Muskan Sonker
Georgia Tech

1:00 PM
Oral 105 - Biomimetic membranes incorporating artificial water channels for high-performance water reverse osmosis desalination
Mihail Barboui (Institut Europeen des Membranes)*

1:20 PM
Oral 106 - Upcycled PVC Support Layer from Waste PVC Pipe for Thin Film Composite Nanofiltration Membranes
Atta Ur Razzaq (The University of Alabama)*; Milad Rabbani Esfahani (The University of Alabama)

1:40 PM
Oral 107 - Maximizing the Wetting Resistance of Fluorine-free Omniphobic Membranes for Hypersaline Wastewater Desalination
Yinchuan Yang (University of British Columbia)*; Sifat Kalam (University of British Columbia); Sadaf Shabanian (University of British Columbia); Kevin Golovin (University of Toronto); Yalei Zhang (Tongji University); Jongho Lee (University of British Columbia)

2:00 PM
Oral 108 - Direct Membrane Osmometry for Rapid Osmotic Pressure Measurements at High Molality
Mi Zhang (University of Connecticut)*; Jeffrey R McCutcheon (University of Connecticut); George Bollas (University of Connecticut); Edward Wazer (University of Connecticut); Wajeha Tauqir (University of Connecticut)

2:20 PM
Oral 109 - XPRIZE Water Scarcity
Hossein Atoufi (XPRIZE)*

2:40 PM
Oral 110 - Delamination Mitigation of Modified Bilayer Membranes using Single Slot Die Roll-to-Roll Manufacturing Process
Minwoo Jung (Georgia Institute of Technology)*; Rachel Kaplan (University of Kentucky); Isabel Escobar (University of Kentucky); Tequila Harris (Georgia Institute of Technology)

NAMS Awards

1:00 PM – 3:00 PM

Room: O'Keeffe

Session Chairs

Xiaobo Dong
Cytiva

Michele Galizia
University of Oklahoma

1:00 PM
Oral 111 - How click chemistry can enable design and investigation of facilitated CO2 separation membranes
Joshua D Moon (University of Florida)*

1:20 PM
Oral 112 - Rational design, synthesis, and characterization of facilitated transport membranes exhibiting enhanced permeability, selectivity, and stability
Matthew T Webb (University of Oklahoma)*; Lucas Condes (University of Oklahoma); Harold Ly (University of Oklahoma); Michele Galizia (University of Oklahoma); Sepideh Razavi (University of Oklahoma)

1:40 PM
Oral 113 - Surface coatings for fouling-resistant membrane distillation
Kecheng Guan (Kobe University)*; Hideto Matsuyama (Kobe University)

2:00 PM
Oral 114 - Ultrahigh Charge Density Ion-Exchange Membranes
David Kitto (University of Michigan)*; José Carlos Díaz (University of Michigan); Carolina Espinoza (University of Michigan); Jacquelyn Zamora (University of Michigan); Jovan Kamcev (University of Michigan)

2:20 PM
Oral 115 - Tuning the selectivity of fouling resistant zwitterionic copolymer membranes
Sam J Lounder (ZwitterCo)*; Ayse Asatekin (Tufts)

2:40 PM
Oral 116 - Upcycling wastewater nitrate into ammonia fertilizer via electrified membrane
Jianan Gao (New Jersey Institute of Technology)*

Advanced Understanding and Synthetic Efforts in Solute Selective Separations

1:00 PM – 3:00 PM

Room: Peralta-Lamy

Session Chairs
DB Bhattacharyya
University of Kentucky

Shashank Kamdar
Dow

1:00 PM
Oral 117 - Microwave dielectric relaxation spectroscopy: A technique to inform ion transport in hydrated polymer membranes
Geoffrey M Geise (University of Virginia)*

1:20 PM
Oral 118 - Insights into methods of determining selectivity of porous ultrafiltration membranes
Suzana Ivandic (UT Austin)*; Mostafa Nassr (UT Austin); Alishba Amjad (UT Austin); Ben Pedretti (MIT); Nathaniel Lynd (The University of Texas at Austin); Lynn Katz (The University of Texas at Austin); Benny Freeman (The University of Texas at Austin)

1:40 PM
Oral 119 - Mixing and Matching: Amphiphilic Polyampholytes and Polyelectrolyte Complexes that Mimic Zwitterionic Behavior for Membrane Applications
Luca Mazzaferro (Tufts University); Ayse Asatekin (Tufts)*

2:00 PM
Oral 120 - Highly rigid and microporous polyimine membranes by interfacial polymerization for efficient hydrocarbon separations
Hiroki Fukuda (University of British Columbia); Jongho Lee (University of British Columbia)*

2:20 PM
Oral 121 - Real-Time, Non-Destructive Monitoring of Supported Liquid Membranes for Biochemical Recovery Using Electrochemical Impedance Spectroscopy
Hiroki Fukuda (University of British Columbia); Jongho Lee (University of British Columbia)*

2:40 PM
Oral 122 - Surface Modification of Nanofiltration Membranes for Selective Nutrient Recovery
Km Prottoy Shariar Piash (West Virginia University)*; Lian-Shin Lin (West Virginia University); Oishi Sanyal (West Virginia University)

Biochemical Applications

1:00 PM – 3:00 PM

Room: Milagro-Kearny

Session Chairs
Hee Jeung Oh
Penn State University

Cristiana Boi
North Carolina State University

1:00 PM
Oral 123 - Membrane Optimization and Filtration Strategies Using Nanoparticles for Enhanced Efficiency in Gene Therapy Applications
Mara Leach (University of Kentukey)*; Gosia Chwatko (University of Kentucky); Dibakar Bhattacharyya (U. Kentucky)

1:20 PM
Oral 124 - Combined Hydrocyclone-Tangential Flow Microfiltration for Bioreactor Harvesting
Amir Hossein Mostafavi (University of Arkansas)*; Liang-Kai Chu (Pennsylvania State University); Xianghong Qian (University of Arkansas); Andrew Zydney (Pennsylvania State University); Ranil Wickramasinghe (University of Arkansas)

1:40 PM
Oral 125 - Charged biosponge polymers for capturing toxic chemotherapy drugs to minimize chemotherapy
Adele Godby (Pennsylvania State University); Rachel Lau (Pennsylvania State University); Hee Jeung Oh (Pennsylvania State University)*

2:00 PM
Oral 126 - Mechanisms of Flux Decline in Crossflow Ultrafiltration of Whey Protein: Comparing Different Models to Experimental Data
Burcu Ozdemir (UCLA)*; Eric Hoek (UCLA)

2:20 PM
Oral 127 - Protocol for purification of circular RNA from a self-splicing reaction by ultrafiltration
Karen d Guillen Cuevas (Clemson University); Marc Birtwistle (Clemson University); Scott Husson (Clemson University)*

2:40 PM
Oral 128 - Patterned anion exchange membrane for improved performance of glucose fuel cells
Thomas A DosSantos DiSorbo (University of Colorado Denver | Anschutz)*

Oral Sessions - Tuesday Evening

Presenting authors are indicated by an asterisk*.

Desalination and Osmotic Processes III

3:30 PM – 5:30 PM

Room: Coronado/DeVargas

Session Chairs
Jovan Kamcev
University of Michigan

Jonathan Maisonneuve
Oakland University

Muskan Sonker
Georgia Tech

3:30 PM
Oral 129 - Permeance Prize: Thin Film Composite Membranes are not Just for Reverse Osmosis Anymore: Recoupling Materials Discovery with Thin-Film Manufacturability
Jeffrey R McCutcheon (University of Connecticut)*

3:50 PM
Oral 130 - Evaluation of Commercial Nanofiltration and Reverse Osmosis Membrane Filtration to Remove PFASs: Effects of Transmembrane Pressures, Water Matrices such as Surfactants, and Ion Valency and Temperatures
Guangyu Zhu (New Jersey Inst of Technology)*; Boris Khusid (New Jersey Inst of Technology); Wen Zhang (New Jersey Institute of Technology); Qingquan Ma (New Jersey Inst of Technology); Neel Ahuja (Millburn High School)

4:10 PM
Oral 131 - A Concise Analytical Model for the System Performance of an RO Desalination Process
Lianfa Song (Texas Tech University)*

4:30 PM
Oral 132 - Novel artificial water channel membranes with improved molecular separation properties in reverse osmosis
Maria Di Vincenzo (KAUST)*; Suzana Nunes (King Abdullah University of Science and Technology (KAUST)); Alberto Tiraferri (Politecnico di Torino)

4:50 PM
Oral 133 - Regulating interfacial polymerization via multi-functional calcium carbonate based interlayer for highly permselective nanofiltration membrane
Mengyang Hu (Kobe University)*; Hideto Matsuyama (Kobe University)

5:10 PM
Oral 134 - Three-dimensional modeling of flow and mass transfer in spacer-filled channels in Reverse osmosis/Nanofiltration processes
Santiago Cespedes (KAUST)*; Fernan David Martinez Jimenez (KAUST); Bastiaan Blankert (KAUST); Cristian Picioreanu (KAUST)

Nams Keynote I

3:30 PM – 5:30 PM

Room: Peralta-Lamy

Session Chairs
Mahdi Malmali
Texas Tech University

Grigorios Panagakos
Carnegie Mellon University

Shouliang Yi
SusMembrane Technologies

3:30 PM
Oral 135 - A Tale of Two Carbon-Capture Properties: Bulk and Thin Films of Mixed Matrix Materials
Haiqing Lin (The State University of New York at Buffalo)*; Gengyi Zhang (The State University of New York at Buffalo); Lingxiang Zhu (National Energy Technology Laboratory)

4:00 PM
Oral 136 - Osmoses: Developing Gas Separation Membranes to Enable Sustainable Energy Solutions

Holden W Lai (Osmoses)*

4:30 PM
Oral 137 - Deep Flue Gas Decarbonization with Novel Membrane Process Designs

Tim Merkel (Membrane Technology & Research Inc.); Trevor Carlisle (Membrane Technology & Research Inc.); Pingjiao Hao (Membrane Technology & Research Inc.); Richard Baker (Membrane Technology & Research Inc.)*

5:00 PM
Oral 138 - Polymer membranes stability under sour-humid natural gas mixtures and their role towards de-carbonization

Nitesh Bhuwania (Chevron)*; Daniel Chinn (Chevron); Thijs Peters (SINTEF); Luca Ansaloni (SINTEF); Oguz Karvan (EMI Twente BV); Alberto Tena (EMI Twente BV); Tymen Visser (EMI Twente BV)

2D Membrane Materials and Processes

3:30 PM – 5:30 PM

Room: Milagro-Kearny

Session Chairs
Reza Foudazi
University of Oklahoma

Ying Labreche
Pall Corporation

3:30 PM
Oral 139 - Strengthening Isoporous Polymer Membranes through Chain Entanglements

Adam N Mann (University of Texas at Austin)*; Benny Freeman (University of Texas at Austin); Gabriel Sanoja (University of Texas at Austin)

3:50 PM
Oral 140 - Highly Selective Thin B-Oriented MFI Zeolite Membranes on Scalable Modified Stainless-Steel Supports

Fateme Banihashemi (Arizona State University); Jerry Lin (Arizona State University)*

4:10 PM
Oral 141 - Ion-sieving membranes based on Polymers of Intrinsic Microporosity for aqueous organic redox flow batteries

Chunchun Ye (School of Chemistry, University of Edinburgh)*; Neil McKeown (School of Chemistry, University of Edinburgh); Qilei Song (Department of Chemical Engineering, Imperial College London); Anqi Wang (Department of Chemical Engineering, Imperial College London); Rui Tan (Warwick Electrochemical Engineering, Warwick Manufacture Group, University of Warwick)

4:30 PM
Oral 142 - Role of salt additives in tuning the structure of β-cyclodextrin-based membranes

Jinting Hu (KU Leuven)*; Sem Bleus (KU Leuven); Linde Achten (KU Leuven); Wim Dehaen (KU Leuven); Xing Yang (KU Leuven)

4:50 PM
Oral 143 - Revealing Molecular Mechanisms in Hierarchical Nanoporous Carbon via Nuclear Magnetic Resonance

Mao, Haiyan *; Halat, David; Pines, Alexander; Cui, Yi; Rimer, Jeffrey

5:10 PM
Oral 144 - Anti- swelling Characteristics Enhanced via Self-Cross-Linking in MXene-Interlaced Graphene Oxide Membranes

Saurabh Kr Tiwary (University Of Houston)*; Alamgir Karim (University of Houston)

Viral and Antibody Filtration

3:30 PM – 5:30 PM

Room: O'Keeffe

Session Chairs

Hee Jeung Oh
Penn State University

Cristiana Boi
North Carolina State University

3:30 PM

Oral 145 - The application of tangential flow filtration for the continuous processing of monoclonal antibodies: process optimization based on particle morphology

Ali Behboudi (The Pennsylvania State University)*; Andrew Zydney (Pennsylvania State University)

3:50 PM

Oral 146 - A more sustainable all-membrane process for the purification of monoclonal antibodies

Cristiana Boi (Università di Bologna)*; Jinxin Fan (North Carolina State University); Joseph Lavoie (North Carolina State University); Ruben Carbonell (North Carolina State University)

4:10 PM

Oral 147 - IgG-BSA Separation and Purification by Internally Staged Ultrafiltration Lixin Feng (New Jersey Institute of Technology); Yufeng Song (New Jersey Institute of Technology); Sagnik Basuray (New Jersey Institute of Technology); Kamalesh K Sirkar (New Jersey Institute of Technology)*

4:30 PM

Oral 148 - Effect of protein on the location of virus entrapment during virus filtration

Wenbo Xu (University of Arkansas); Hironobu Shirataki (Asahi Kasei Corporation); Xianghong Qian (University of Arkansas); Ranil Wickramasinghe (University of Arkansas)*

4:50 PM

Oral 149 - The Effects of Pressure and Process Interruption on The Location of Virus Particles Captured during Virus Filtration

Wenbo Xu (University of Arkansas)*; Hironobu Shirataki (Asahi-Kasei); Ranil Wickramasinghe (University of Arkansas); Xianghong Qian (University of Arkansas)

5:10 PM

Oral 150 - Bioseparation Membranes from Crosslinking of Pluronic F127 Self-assembly

Tabatabaei, Seyed Mostafa (University of Oklahoma)*; Foudazi, Reza Mostafa (University of Oklahoma)

Oral Sessions - Wednesday Morning

Presenting authors are indicated by an asterisk*.

Membranes for Carbon Capture II

8:30 AM – 10:50 AM

Room: Milagro-Kearny

Session Chairs

David Hopkinson
National Energy Technology Laboratory

Joshua D. Moon
University of Florida

Zhongyun Liu
Georgia Tech

8:30 AM
Oral 151 - Large pilot testing of MTR's membrane-based post-combustion CO2 capture process

Richard Baker (Membrane Technology & Research Inc.); Phillip Behm (Membrane Technology & Research Inc.); Alicia Breen (Membrane Technology & Research Inc.); Brice Freeman (Membrane Technology & Research Inc.); Pingjiao Hao (Membrane Technology & Research Inc.); Thomas Hofmann (Membrane Technology& Research Inc.); Jay Kniep (Membrane Technology & Research Inc.); Tim Merkel (Membrane Technology & Research Inc.); Witopo Salim (Membrane Technology & Research Inc.)*; Ray McKaskle (Trimeric Corporation); Dana Pierik (Sargent & Lundy); William Morris (Wyoming Integrated Test Center)

8:50 AM
Oral 152 - Designing membranes using bottlebrush poly(1,3-dioxolane) acetate for CO2 capture

Fathy Attia (The State University of New York at Buffalo)*; Thien N Tran (University at Buffalo, The State University of New York); Erda Deng (State University of New York at Buffalo); Bhanuprakash Valluri (University at Buffalo); Liang Huang (The State University of New York at Buffalo); Haiqing Lin (The State University of New York at Buffalo)

9:10 AM
Oral 153 - Random Poly(1,3-dioxolane)-based Terpolymers for CO2 Separations

Justin J Rosenthal (The University of Texas at Austin)*; Isaac Tan (The University of Texas at Austin); Julian Richardson (University of Texas, Austin); Nathaniel Lynd (The University of Texas at Austin); Benny Freeman (The University of Texas at Austin)

9:30 AM
Oral 154 - Highly Permeable Rubbery Thin Film Composite Membranes for CO2 Capture from Steel Mills

Lingxiang Zhu (National Energy Technology Laboratory)*; Thien Tran (National Energy Technology Laboratory); Fangming Xiang (NETL); Victor A Kusuma (NETL); Cheick Dosso (Carnegie Mellon University); Hector A Pedrozo (Carnegie Mellon University); Grigorios Panagakos (Carnegie Mellon University); Neil Pergar (U. S. Steel); Brenda Petrilena (U. S. Steel); David Hopkinson (NETL)

9:50 AM
Oral 155 - Investigating PIM-1 and cPIM-1 applications for CCUS under real-world conditions

Sean R McIntyre (surface measurement systems)*; Andrew Foster (University of Manchester); Peter M. Budd (University of Manchester); Paul Iacomi (Surface Measurement Systems); Daryl Williams (Imperial College London)

10:10 AM
Oral 156 - Retarded O2 transport in Co2+-coordinated supramolecular polymer networks for membrane CO2/O2 separations

Haiqing Lin*

10:30 AM

Oral 157 - A Membrane Contactor Enabling Energy-Efficient CO2 Capture from Point Sources with Physical Solvent

Nicholas Gregorich (Oak Ridge National Laboratory)*; Mary Irwin (Oak Ridge National Laboratory); Zachary Coin (Oak Ridge National Laboratory); Stephen JA DeWitt (Arkema Inc); Ramesh Bhave (Ornl); David Sholl (Oak Ridge National Laboratory); Syed Islam (Oak Ridge National Laboratory)

Specific Ion Separations

8:30 AM – 10:50 AM

Room: Peralta-Lamy

Session Chairs

Ngoc Bui
University of Oklahoma

Geoffrey Geise
University of Virginia

8:30 AM

Oral 158 - Elucidating selectivity mechanisms of anionic pollutants in nanofiltration membranes

Idit Shefer (Technion)*; Razi Epsztein (Technion)

8:50 AM

Oral 159 - A study of NF membrane assessment as pretreatment for lithium recovery from salt-lake brine

Juan Zhai (Texas Tech University)*; Yuexiao Shen (Texas Tech University); Mahdi Malmali (Texas Tech University); Rajesh Khare (Texas Tech University); Amrika Deonarine (Texas Tech University); Shubhra Bhattacharjee (Texas Tech University); Adegbola Balogun (Texas Tech University)

9:10 AM

Oral 160 - Understanding the dopant effects on the selective transport of cations in conductive polymers

Meng Shen (California State University, Fullerton)*

9:30 AM

Oral 161 - Ion-selective Membranes for Critical Materials Separation

Ngoc Bui (The University of Oklahoma)*

9:50 AM

Oral 162 - Aligned macrocycle pores in ultrathin films for accurate molecular sieving

Zhiwei Jiang (Queen Mary University of London)*; Ruijiao Dong (Shanghai Jiao Tong University); Austin Evans (University of Florida); Niklas Biere (Bielefeld University); Mahmood Abdulsalam (Imperial College London); siyao li (Imperial College London); Dario Anselmetti (Bielefeld University); William Dichtel (Northwestern University); Andrew Livingston (Queen Mary University of London)

10:10 AM

Oral 163 - Tailoring Interlayer Spacing and Surface Chemistry of MXene Free-standing Membranes for Selective Separation of Precious Ions from Water

Ahmed Al Mayyahi (the University of Alabama)*; Milad Rabbani Esfahani (The University of Alabama)

10:30 AM

Oral 164 – Selective ion conducting polymers for non-aqueous redox flow battery application

Patrick McCormack (UVA), Gary Koenig (UVA), Geoffrey Geise (UVA)*

Scaling Up: Bridging the Gap Between Lab and Industry

8:30 AM – 10:50 AM

Room: O'Keeffe

Session Chairs

Oguz Karvan
Twente University

Guy Ramon
Isreal Institute of Technology

8:30 AM
Oral 165 - Electrosprayed membranes for CO2-selective gas separations

Noah Ferguson (University of Connecticut)*; Arjun DeJesus (University of Connecticut); Srivatsa Bettahalli (University of Connecticut); Jeffrey R McCutcheon (University of Connecticut)

8:50 AM
Oral 166 - Simplified Synthesis Methods for Membrane-Protein Nanosheets and Biomimetic Membranes

Ronald J Vogler (The University of Texas at Austin)*; Katherine Kimball (The University of Texas at Austin); Dominic Bujanos (The University of Texas at Austin); Raman Dhiman (The University of Texas at Austin); Benny Freeman (The University of Texas at Austin); Manish Kumar (University of Texas at Austin)

9:10 AM
Oral 167 - Evaluation of spacer-induced hydrodynamic mixing using particle image velocimetry: Impact on membrane distillation performance

Hadis Matinpour (Texas Tech University); Mahdi Malmali (Texas Tech University)*; Bosong Lin (Texas Tech University)

9:30 AM
Oral 168 - A Fractal Analysis Based Metrology to Quantify Porous Structure of Membranes

Riccardo Lovison (University of Colorado Boulder)*; Yifu Ding (Membrane Science, Engineering, and Technology (MAST) Center, Department of Mechanical Engineering, University of Colorado Boulder); Christina Carbrello (MilliporeSigma)

9:50 AM
Oral 169 - Design of Experiments to Optimize the Polyimide Membrane Structure

Margaret Lee (Pall Corporation)*; Siavash Darvishmanesh (Pall corporation)

10:10 AM
Oral 170 - Spiral Wound Module Development and Scale-up for Olefin-Paraffin Separations with OptipermTM Membrane System

Jacob Wargo (Compact Membrane Systems)*; Christine Parrish (Compact Membrane Systems); Brandon J Burghardt (Compact Membrane Systems); Bill Charlton (Compact Membrane Systems)

10:30 AM
Oral 171 - Effect of polymer molecular weight on the long-term process stability of crosslinked polybenzimidazole (PBI) membranes

Adam Oxley (Exactmer Ltd)*; Andrew Livingston (Queen Mary University of London)

Contaminants Removal from Water I

8:30 AM – 10:50 AM

Room: Coronado/DeVargas

Session Chairs

Zora Huang
Chevron

Jian Xu
Bayer

Ahmad Shamsabadi
Pall Corporation

8:30 AM
Oral 172 - Structure-property relationships in intercalated graphene oxide nanofiltration membranes for enhanced separation of salts and organics from aqueous streams

Muskan Sonker (Georgia Institute of Technology)*; Sankar Nair (Georgia Institute of Technology); Kaung Su Khin Zaw (Georgia Institute of Technology); Hazel P. Dhruve (, Georgia Institute of Technology); Mahsa Abbaszadeh (Albemarle); Madeline P. Garell (Georgia Institute of Technology); Michael B. Salerno (Phillips 66 Research Center); Marta Hatzell (Georgia Institute of Technology); Meisha L. Shofner (Georgia Institute of Technology)

8:50 AM
Oral 173 - Removal of Dodecyl Benzenesulfonic Acid by Amine-Modified Membrane Adsorbers

Hadi Rouhi (University of Alabama)*; Nick Vacca (University of Alabama); Jess Moskey (University of Alabama); Steven T Weinman (The University of Alabama)

9:10 AM
Oral 174 - Thin Film Nanocomposite Membranes Embedded with ZIF-67 for Desalination and Boron Removal

Suer Kurklu Kocaoglu (The University of Texas at Austin)*; Benny Freeman (The University of Texas at Austin); S. Birgül Tantekin-Ersolmaz (Istanbul Technical University)

9:30 AM
Oral 175 - Unlocking the power of a gutter layer: A new frontier in dye-desalination using amine-crosslinked nanoporous graphene oxide (aHGO) platform

Ameya Tandel (State University of New York, Buffalo, NY)*; Haiqing Lin (The State University of New York at Buffalo)

9:50 AM
Oral 176 - Experimental and Modeling Results for the Rejection of Ultrashort Per-and Polyfluoroalkyl Substances

Aron Griffin (Colorado School of Mines)*; Timothy Strathmann (Colorado School of Mines); Christopher Bellona (Colorado School of Mines)

10:10 AM

Oral 177 - Polymer evolution and fabrication of PVDF membranes for contaminant removal from water

Kaplan, Rachel (University of Kentucky)*; Escobar, Isabel (University of Kentucky)

10:30 AM
Oral 178 - Zr-based MOF-NF-Membrane for Selective Removal of Selenium from Produced water

Sweta A Modak (The University of Alabama)*; Milad Rabbani Esfahani (The University of Alabama)

Oral Sessions - Wednesday Afternoon

Presenting authors are indicated by an asterisk*.

Mixed Matrix and Composite Membranes for Gas Separation

1:00 PM – 3:00 PM

Room: Milagro-Kearny

Session Chairs

Shailesh Singh Dangwal
Oak Ridge National Laboratory

Jay (Junqiang) Liu
Dow

Hani A.E. Hawa
Precision Combustion, Inc.

1:00 PM
Oral 179 - UV-Crosslinked Nanocomposite Membranes for Fast and Selective Transport of Small Molecules

Joo-Eon Kim (Korea Research Institute of Chemical Technology (KRICT))*; Mi-Hee Ryu (Korea Research Institute of Chemical Technology (KRICT)); Ahrumi Park (Korea Research Institute of Chemical Technology (KRICT)); Jaesung Park (Korea Research Institute of Chemical Technology (KRICT))

1:20 PM
Oral 180 - Computational insights from the polymer/MOF interface in Mixed-Matrix Membranes

Moein Hajian Zaker Damghan (Texas Tech University)*; Hyunhee Lee (MIT); Zachary Smith (MIT); Joshua Howe (Texas Tech University)

1:40 PM
Oral 181 - Leveraging anomalous gas transport in thin-film composite membranes based on polymers of intrinsic microporosity

Taehoon Lee (MIT, ChemE)*; Ho Bum Park (Hanyang University); Zachary Smith (MIT)

2:00 PM
Oral 182 - CO2/C2H2 Transport in SSZ-13 Zeolite Crystals and Membranes, and the Implication on Zeolite Membrane Microstructure

Ninad Anjikar (Cleveland State University); Kevin Hinkle (University of Dayton); Orhan Talu (Cleveland State University); Qiang Fu (Georgia Institute of Technology); Sankar Nair (Georgia Institute of Technology); Shaowei Yang (Cleveland State University)*

2:20 PM
Oral 183 - Gas Permeation and Separation Characteristics of Microporous TpHz COF Membranes Synthesized by Substrate-Assisted Interfacial Polymerization

Jose E Lopez-Cazares (Arizona State University)*; Jerry Lin (Arizona State University); Kailong Jin (Arizona State University); Kyungtae Kim (Los Alamos National Lab)

2:40 PM
Oral 184 - Towards the Synthesis of Mixed Matrix Membranes Using Vapor Phase Infiltration

Seancarlos Gonzalez (University of Washington)*; Sarah West (University of Washington); Yuri Choe (University of Washington); Kellen McKinney (University of Washington); Cecilia Osburn (University of Washington); David S Bergsman (University of Washington)

Selective Separations Modeling

1:00 PM – 3:00 PM

Room: O'Keeffe

Session Chairs

Meng Shen
California State University, Fullerton

Jianjia Yu
New Mexico Tech

1:00 PM
Oral 185 - Ion and water dynamics in the transition from dry to wet conditions in salt-doped PEG

Nico Marioni (The University of Texas at Austin)*; Oscar Nordness (University of California Santa Barbara); Zidan Zhang (The University of Texas at Austin); Rahul Sujanani (University of California Santa Barbara); Benny Freeman (The University of Texas at Austin); Rachel Segalman (University of California Santa Barbara); Raphaële Clément (University of California Santa Barbara); Venkat Ganesan (The University of Texas at Austin)

1:20 PM
Oral 186 - The influence of Donnan – Manning and dielectric exclusion on ion partitioning in sulfonated polysulfone

Sean M Bannon (University of Virginia)*; Geoffrey M Geise (University of Virginia)

1:40 PM
Oral 187 - Influence of backbone rigidity and water content on lithium/magnesium selectivity in hydrated polymer membranes

Paul R Irving (University of Texas at Austin)*; Nico Marioni (The University of Texas at Austin); Harnoor Sachar (University of Texas at Austin); Everett S Zofchak (The University of Texas at Austin); Kevin Reimund (University of Texas at Austin); Zidan Zhang (University of Texas at Austin); Tyler J Duncan (The University of Texas at Austin); Benny Freeman (The University of Texas at Austin); Venkat Ganesan (University of Texas at Austin)

2:00 PM
Oral 188 - Molecular Dynamics Simulations of Lithium/Magnesium Mixed Solutions in Polyamide Membranes

Richard Lueptow (Northwestern University)*; Sinan Keten (Northwestern University); Suwei Liu (Northwestern University)

2:20 PM
Oral 189 - Modeling polymer aging in solvent environments with symbolic regression

Matthew Rivera (MIT)*; Zachary Smith (MIT)

2:40 PM
Oral 190 - Ion-ion Separations in Biomimetic Water Channels

Tyler J Duncan (The University of Texas at Austin)*; Harekrushna Behera (The University of Texas at Austin); Nico Marioni (The University of Texas at Austin); Meron Tadesse (University of Texas Austin); Paul R Irving (University of Texas at Austin); Everett S Zofchak (The University of Texas at Austin); Manish Kumar (University of Texas at Austin); Venkat Ganesan (University of Texas Austin)

NAMS Keynote II

1:00 PM – 3:00 PM

Room: Peralta-Lamy

Session Chairs

Mahdi Malmali
Texas Tech University

Grigorios Panagakos
Carnegie Mellon University

Shouliang Yi
SusMembrane Technologies

1:00 PM
Oral 191 - Dirt and waste as source materials for resource recovery membranes

Seth Darling (Argonne National Laboratory)*

1:30 PM
Oral 192 - Pall's Filtration Solution for Separation and Purification in Enabling Decarbonization
Varun Vakharia (Pall Corporation)*

2:00 PM
Oral 193 - Opportunities and challenges for CO2 separation membranes in decarbonization of the electric power, cement, and steel industries
David Hopkinson (NETL)*; Lingxiang Zhu (National Energy Technology Laboratory); Thien Tran (National Energy Technology Laboratory); Victor A Kusuma (NETL); Eric Grol (NETL); Timothy Fout (NETL); Grigorios Panagakos (Carnegie Mellon University); Glenn Lipscomb (University of Toledo, Toledo); Brenda Petrilena (U. S. Steel); Neil Pergar (U. S. Steel)

2:30 PM
Oral 194 - Carbon capture through membranes: leveraging multiphysics modeling, dimensional analysis and machine learning to scale up and optimize devices and processes for decarbonization
Hector A Pedrozo (Carnegie Mellon University); Cheick Dosso (Carnegie Mellon University); Thien Tran (National Energy Technology Laboratory); Lingxiang Zhu (National Energy Technology Laboratory); Victor A Kusuma (NETL); David Hopkinson (NETL); Lorenz Biegler (Carnegie Mellon University); Grigorios Panagakos (Carnegie Mellon University)*

Contaminants Removal from Water II

1:00 PM – 3:00 PM

Room: Coronado/DeVargas

Session Chairs
Zora Huang
Chevron

Ahmad Shamsabadi
Pall Corporation

1:00 PM
Oral 195 - Real-Time Induced Magnetic Vibrational Based Antifouling Mechanism for Ultrafiltration (UF) membrane
Jasneet K Pala (The University of Alabama)*; Milad Rabbani Esfahani (The University of Alabama); Ryan Tracy (The University of Alabama); Nima Mahmoodi (The University of Alabama)

1:20 PM
Oral 196 - Catalytic Membranes Based on Reduced Graphene Oxide for Micropollutant Treatment Enabled by Peracetic Acid
Erda Deng (State University of New York at Buffalo)*; Siavash Mohamadi (State University of New York at Buffalo); Zachary Kralles (State University of New York at Buffalo); Ning Dai (University at Buffalo); Haiqing Lin (The State University of New York at Buffalo)

1:40 PM
Oral 197 - Designing the fluorinated superfine-activated carbon adsorptive membrane for selective removal of long- and short-chain PFAS
Medha Kasula (The University of Alabama)*; Milad Rabbani Esfahani (The University of Alabama)

2:00 PM
Oral 198 - Dynamic ammonium retention for nutrient separation from manure digestate
Marrit van der Wal (Eindhoven University of Technology)*; Zandrie Borneman (Eindhoven University of Technology); Kitty Nijmeijer (Eijndhoven)

2:20 PM

Oral 199 - Advanced functionalized polyether ether ketone/polysulfone membranes with 2D nanomaterials for enhanced water treatment performance

Muhammad Usman Yousaf (University of Kentucky)*; Nerissa Kiprotich (University of Kentucky); Anika Azme (University of Nebraska, Lincoln); Lucca Madeo Cortarelli (University of Kentucky); Nirupam Aich (University of Nebraska, Lincoln); Olga Tsyusko (University of Kentucky); Isabel Escobar (University of Kentucky)

2:40 PM

Oral 200 - Distributed Membrane-Based Water Desalination and Purification Systems: Automation, Data-Driven Operational Models, and Performance Forecasting

Nora Marki (University of California, Los Angeles) *, Yoram Cohen (University of California, Los Angeles), Yang Zhou (Shanghai University), Yakubu Jarma (University of California, Los Angeles), Maria Soto (University of California, Los Angeles), Christian Aguilar (University of California, Los Angeles), Phoebe Strauss, Muhammad Bilal (University of California, Los Angeles)

Poster Sessions - Monday

Presenting authors are indicated by an asterisk*.

Poster Session (Sweeney A-C)

7:30 PM – 10:00 PM

Session Chairs

Yuexiao Shen
Texas Tech University

Swatantra P. Singh
Indian Institute of Technology Bombay

Poster 1 - Investigating PIM-1 and cPIM-1 applications for CCUS under real-world conditions

Sean R McIntyre (surface measurement systems)*; Andrew Foster (University of Manchester); Peter M. Budd (University of Manchester); Paul Iacomi (Surface Measurement Systems); Daryl Williams (Imperial College London)

Poster 2 - Exploring the separation efficiency of ultra-selective 3D CANAL hydrocarbon polymeric membranes for hydrogen-based separations after long-term aging

Jing Ying Yeo (MIT)*; Francesco Maria Benedetti (Massachusetts Institute of Technology); Zachary Smith (MIT)

Poster 3 - Engineering SPEEK-based membranes for high-performance air dehumidification

Faheem Hassan Akhtar (LUMS)*

Poster 4 - Understanding the mechanism of additive-assisted isoporous membrane formation through X-ray scattering

Noah P Wamble (University of Texas at Austin)*; Mostafa Nassr (UT Austin); Matthew Landsman (Lawrence Berkeley National Lab); Louise Kuehster (University of Texas at Austin); Joan Brennecke (The University of Texas at Austin); Nathaniel Lynd (The University of Texas at Austin); Gregory Su (Lawrence Berkeley National Lab); Benny Freeman (The University of Texas at Austin)

Poster 5 - Impact of Diverse Crosslinkers on Salt Transport in Phenyl Acrylate-Based Anion Exchange Membranes for Artificial Photosynthesis

Yihung Lin (Auburn University)*

Poster 6 - Ion and water dynamics in the transition from dry to wet conditions in salt-doped PEG

Nico Marioni (The University of Texas at Austin)*; Oscar Nordness (University of California Santa Barbara); Zidan Zhang (The University of Texas at Austin); Rahul Sujanani (University of California Santa Barbara); Benny Freeman (The University of Texas at Austin); Rachel Segalman (University of California Santa Barbara); Raphaële Clément (University of California Santa Barbara); Venkat Ganesan (The University of Texas at Austin)

Poster 7 - Ultra-high hydrogen/carbon dioxide selectivity in carbon hollow fiber membranes

Ching-En Ku (University of Maryland); Gaurav Iyer (University of Maryland); Chen Zhang (University of Maryland)*

Poster 8 - Synthesis and Testing of Electrochemical Activity in a Polyether Sulfone-Based Ultrafiltration Membrane for Contaminant Removal

Utkarsh Misra (IIT Bombay)*; Kritika Jashrapuria (Indian Institute of Technology Bombay); Swatantra Pratap Singh (IIT BOMBAY)

Poster 9 - Membrane Optimization and Filtration Strategies Using Nanoparticles for Enhanced Efficiency in Gene Therapy Applications

Mara Leach (University of Kentukey)*; Gosia Chwatko (University of Kentucky); Dibakar Bhattacharyya (U. Kentucky)

Poster 10 - Optimizing electrospun composite-fiber patterns on nanofiltration membranes for improved biofouling resistance

Amos Taiswa (Montana Tech)*; Jessica Andriolo (Montana Tech); Jack Skinner (Montana Tech)

Poster 11 - A Click Chemistry-Based Platform for the Selective Separation of Lanthanides from Industrial Waste

Lianna Johnson (CWRU)*; Bernadette Schneider (CWRU); Christine E Duval (Case Western Reserve University)

Poster 12 - Strategy to improve permeate flux for typical nanofiltration membranes by modifying the polysulfone support and their application in desalination and micropollutant removal.

Elham Saud Alkhulaify (King Fahd University of Petroleum and Minerals); Shehzada Muhammad Sajid Jillani (King Fahd University of Petroleum and

Minerals)*; Khalid Alhooshani (King Fahd University of Petroleum and Minerals)

Poster 13 - Internal 3-dimensional structure of hydrated polyamide RO membranes revealed by cryo-electron tomography (cryo-ET)
Chenhao Yao (University of Texas at Austin)*

Poster 14 - Synthesis and Characterization of ZnCl2-Immobilized Molten Salt (IMS) membrane for NH3 separation
Moses T Adejumo (Clarkson University)*; Simona Liguori (Clarkson University)

Poster 15 - Co-Transport of Alcohols and Carboxylate Ions in Sulfonated Ion Exchange Membranes with PEGPEA as a Blocking Group
Pravin Parasakthi Aravindhan (Auburn University)*; Bryan S Beckingham (Auburn University); Jung Min Kim (University of Delaware)

Poster 16 - Mixed counter-ion sorption in ion-exchange membranes: understanding the role of membrane water content
Harsh Patel (University of Michigan)*; José Carlos Díaz (University of Michigan); Jovan Kamcev (University of Michigan)

Poster 17 - Removal of Dodecyl Benzenesulfonic Acid by Amine-Modified Membrane Adsorbers
Hadi Rouhi (University of Alabama)*; Nick Vacca (University of Alabama); Jess Moskey (University of Alabama); Steven T Weinman (The University of Alabama)

Poster 18 - Amine Functionalized Supported Ionic Liquid Membranes (SILMs) for CO2/N2 Separation
Antoine Chamoun-Farah (University of Texas at Austin)*; Austin Keller (University of Texas at Austin); Mariam Balogun (University of Texas at Austin); Louise Canada (University of Texas at Austin); Joan Brennecke (The University of Texas at Austin); Benny Freeman (The University of Texas at Austin)

Poster 19 - Surfactant-assisted interfacial polymerization of polyamide reverse osmosis membranes for desalination applications

Vrajkumar Shah (The University of Alabama)*; Bryn Larson (The University of Alabama); Steven Weinman (The University of Alabama)

Poster 20 - Revisiting experimental techniques and theoretical models for estimating the solubility parameter of rubbery and glassy polymer
Matthew T Webb (University of Oklahoma)*; Lucas Condes (University of Oklahoma); William Box (University of Oklahoma); Harold Ly (University of Oklahoma); Sepideh Razavi (University of Oklahoma); Michele Galizia (University of Oklahoma)

Poster 21 - Three-dimensional salt and water transport model in composite membranes: effect of the internal structure.
Fernan David Martinez Jimenez (KAUST)*

Poster 22 - Dynamic ammonium retention for nutrient separation from manure digestate
Marrit van der Wal (Eindhoven University of Technology)*; Zandrie Borneman (Eindhoven University of Technology); Kitty Nijmeijer (Eijndhoven)

Poster 23 - Enhanced bipolar membrane water disassociation reaction with layer-by-layer assembled clay composite junctions
Nadia Boulif (Eindhoven University of Technology)*; Menno Houben (Eindhoven University of Technology); Zandrie Borneman (Eindhoven University of Technology); Kitty Nijmeijer (Eijndhoven)

Poster 24 - Amine Modified PIM-1 for Highly Selective Olefin/Paraffin Separation
Bo Wei Cynthia Chen (University of Notre Dame)*; Casey O'Brien (University of Notre Dame)

Poster 25 - Hydrothermally Stable Polyorganosilica Membranes Synthesized by Oxygen Plasma Treatment of Phenyl-Containing Polysiloxanes for H2 /CO2 Separation
Vinh Bui (University at Buffalo)*; Varun Reddy Satti (University at Buffalo); Elizabeth Haddad (University at Buffalo); Ameya Tandel (State University of New York, Buffalo, NY); Narjes Esmaeili (The State University of New York at Buffalo); Sai Srikar Chundury (University at Buffalo); Lingxiang Zhu (National Energy Technology Laboratory); Haiqing Lin (The State University of New York at Buffalo)

Poster 26 - Towards the Synthesis of Hydrophilic Membranes for PFAS Removal by Molecular Layer Deposition
Joelle V Scott (University of Washington)*; Jocelyne Booth (University of Washington); David S Bergsman (University of Washington)

Poster 27 - Co-Ion Specific Effect on Li Salt Sorption in Cation Exchange Membranes (CEM)
Rachel Z Huang (UT Austin)*; Lynn Katz (The University of Texas at Austin); Benny Freeman (The University of Texas at Austin)

Poster 28 - Membrane-Based Dehumidification using Fertilizer Liquid Desiccants: Experimental and Numerical Evaluation of Energy Efficiency
Sandeep Aryal (Oakland University)*; Sarah Moussaddy (Oakland University); Jonathan Maisonneuve (Oakland University)

Poster 29 - Influence of backbone rigidity and water content on lithium/magnesium selectivity in hydrated polymer membranes
Paul R Irving (University of Texas at Austin)*; Nico Marioni (The University of Texas at Austin); Harnoor Sachar (University of Texas at Austin); Everett S Zofchak (The University of Texas at Austin); Kevin Reimund (University of Texas at Austin); Zidan Zhang (University of Texas at Austin); Tyler J Duncan (The University of Texas at Austin); Benny Freeman (The University of Texas at Austin); Venkat Ganesan (University of Texas at Austin)

Poster 30 - Distributed Membrane-Based Water Desalination and Purification Systems: Automation, Data-Driven Operational Models, and Performance Forecasting
Yoram Cohen (Professor)*; Nora Marki (University of California, Los Angeles); Yang Zhou (Shanghai University); Yakubu Jarma (University of California, Los Angeles); Maria Soto (University of California, Los Angeles); Christian Aguilar (University of California, Los Angeles); Phoebe Strauss (University of California, Los Angeles); Muhammad Bilal (University of California, Los Angeles)

Poster 31 - Oxidative, catalytic oxidative and non-oxidative electro-cleaning of fouled conductive membranes
Najmul Haque Barbhuiya (Indian Institute of Technology Bombay)*; Swatantra Pratap Singh (IIT BOMBAY)

Poster 32 - Understanding the influence of external sodium chloride concentration on ion diffusion in sulfonated polymers
Sean M Bannon (University of Virginia)*; Ethan Kutner (University of Virginia); Ben Garretson (University of Virginia); Geoffrey M Geise (University of Virginia)

Poster 33 - Evaluation of air addition into methane dry reforming reaction in Pd/YSZ membrane reactor for syngas production
Omid Jazani (Clarkson University)*; Simona Liguori (Clarkson University)

Poster 34 - Techno-economic feasibility analysis of membrane-based purification platforms for AAV vectors production
Juan J Romero (Clemson University)*; Eleanor Jenkins (Clemson University); Jacob Monroe (University of Arkansas); Ranil Wickramasinghe (University of Arkansas); Xianghong Qian (University of Arkansas); Dibakar Bhattacharyya (U. Kentucky); Scott Husson (Clemson University)

Poster 35 - Developments in the Preparation of PVDF Hollow Fiber Membranes from Alternate Solvents
Stephen JA DeWitt (Arkema Inc)*

Poster 36 - A study of NF membrane assessment as pretreatment for lithium recovery from salt-lake brine
Juan Zhai (Texas Tech University)*; Yuexiao Shen (Texas Tech University); Mahdi Malmali (Texas Tech University); Rajesh Khare (Texas Tech University); Amrika Deonarine (Texas Tech University); Shubhra Bhattacharjee (Texas Tech University); Adegbola Balogun (Texas Tech University)

Poster 37 - Volatile fatty acid selectivity during electrodialysis: role of solution composition on volatile fatty acid mobility and partitioning in an anion exchange membrane
Holly M Haflich (University of North Carolina- Chapel Hill)*; Joshua Singleton (University of North Carolina at Chapel Hill); Orlando Coronell (University of North Carolina at Chapel Hill)

Poster 38 - Effect of Pattern Size Relative to Oil Droplet Size on Oil Fouling in Nanofiltration Poster
Lauren M Ward (The University of Alabama)*; Catherine Martin (The University of Alabama); Steven T Weinman (The University of Alabama)

Poster 39 - Unusual transport of phenolic compounds in polymer membranes.
Woo Jin Jang (Georgia Institute of Technology)*

Poster 40 - Organic solvent nanofiltration and reverse osmosis performance of carboxylated

PIM-1 in neat and inter-crosslinked mixed membrane configurations
Lucas Condes (University of Oklahoma)*; Matthew T Webb (University of Oklahoma); Will Box (University of Oklahoma); Michele Galizia (University of Oklahoma)

Poster 41 - Combined Hydrocyclone-Tangential Flow Microfiltration for Bioreactor Harvesting
Amir Hossein Mostafavi (University of Arkansas)*; Liang-Kai Chu (Pennsylvania State University); Xianghong Qian (University of Arkansas); AAA ZZZ (Pennsylvania State University); Ranil Wickramasinghe (University of Arkansas)

Poster 42 - UV-Crosslinked Nanocomposite Membranes for Fast and Selective Transport of Small Molecules
Joo-Eon Kim (Korea Research Institute of Chemical Technology (KRICT))*; Mi-Hee Ryu (Korea Research Institute of Chemical Technology (KRICT)); Ahrumi Park (Korea Research Institute of Chemical Technology (KRICT)); Jaesung Park (Korea Research Institute of Chemical Technology (KRICT))

Poster 43 - Optimization of Fouling Resistant Polyelectrolyte Coacervate Membranes via Aqueous Phase Separation
Shao-Hsiang Hung (University of Massachusetts Amherst)*; Jessica Schiffman (University of Massachusetts Amherst)

Poster 44 - Scalable Fabrication of Antifouling Graphite-based Electrically Conductive Membranes for Water Treatment
Wei Zhang (University of North Carolina at Chapel Hill)*; Nick Guan Pin Chew (University of North Carolina at Chapel Hill); Orlando Coronell (University of North Carolina at Chapel Hill)

Poster 45 - Polymer blending to overcome permeability/selectivity tradeoff in polybenzimidazole for H2/CO2 separation
Narjes Esmaeili (The State University of New York at Buffalo)*; LEIQING HU (University at Buffalo); Erda Deng (State University of New York at Buffalo); Haiqing Lin (The State University of New York at Buffalo)

Poster 46 - Understanding Polyfluoroalkyl Substance Selectivity in Anion Exchange Membrane Systems
Sam S Thompson (University of Kentucky)*; Angela Gutierrez (University of Kentucky); Rollie Mills (U. Kentucky); Dibakar Bhattacharyya (U. Kentucky)

Poster 47 - Developments in Diffusive, Scale-able, and Flexible Graphene Membrane Electrodes
Caleb Song (University of Colorado Boulder)*; Thomas A DosSantos DiSorbo (University of Colorado Denver | Anschutz)

Poster 48 - Thermomechanical Bonding Between Polypropylene and Polyethersulfone Membranes
Rajan Jain (Membrane Science, Engineering, and Technology (MAST) Center, Department of Mechanical Engineering, University of Colorado Boulder)*; Jaylene Martinez (CU Boulder Mechanical Engineering); Christina Carbrello (MilliporeSigma); Kathy Youngbear (MilliporeSigma); Sean Foley (MilliporeSigma); Rong Long (Membrane Science, Engineering, and Technology (MAST) Center, Department of Mechanical Engineering, University of Colorado Boulder); Yifu Ding (Membrane Science, Engineering, and Technology (MAST) Center, Department of Mechanical Engineering, University of Colorado Boulder)

Poster 49 - Acoustic microstreaming bubble-trapping-structures with chemically pinned bubble oscillations for continuous membrane fouling mitigation
Kieran R Fung (Membrane Science, Engineering, and Technology (MAST) Center, Department of Mechanical Engineering, University of Colorado Boulder)*; Yifu Ding (Membrane Science, Engineering, and Technology (MAST) Center, Department of Mechanical Engineering, University of Colorado Boulder); Xiaoyun Ding (Membrane Science, Engineering, and Technology (MAST) Center, Department of Mechanical Engineering, University of Colorado Boulder); Jason P. Killgore (Applied Chemicals and Materials Division, National Institute of Standards and Technology, Boulder, CO); Thomas Kolibaba (Applied Chemicals and Materials Division, National Institute of Standards and Technology, Boulder, CO); Benjamin Caplins (Applied Chemicals and Materials Division, National Institute of Standards and Technology, Boulder, CO)

Poster 50 - Random Poly(1,3-dioxolane)-based Terpolymers for CO2 Separations
Justin J Rosenthal (The University of Texas at Austin)*; Isaac Tan (The University of Texas at Austin); Julian Richardson (University of Texas, Austin); Nathaniel Lynd (The University of Texas at Austin); Benny Freeman (The University of Texas at Austin)

Poster 51 - High Recovery Desalination of High Mineral Scaling Source Water via Integrated RO-NF Concentrator, Concentrate Desupersaturation and Second Stage RO-NF Desalting.
Yakubu Jarma (University of California, Los Angeles); Jose Morales (University of California, Los Angeles); Phoebe Strauss (University of California, Los Angeles); Yoram Cohen (Professor)*

Poster 52 - An all-polyamide organic solvent nanofiltration membrane
Cheng Peng (New Jersey Institute of Technology); Mahesh Patil (New Jersey Institute of Technology); Yufeng Song (New Jersey Institute of Technology); John Chau (New Jersey Institute of Technology); Kamalesh K Sirkar (New Jersey Institute of Technology)*

Poster 53 - High-Throughput Experimentation for High-Pressure Reverse-Osmosis Membranes
Hongchen Wang (University of Toronto)*

Poster 54 - Evaluation of Commercial Nanofiltration and Reverse Osmosis Membrane Filtration to Remove PFASs: Effects of Transmembrane Pressures, Water Matrices such as Surfactants, and Ion Valency and Temperatures
Guangyu Zhu (New Jersey Inst of Technology)*; Boris Khusid (New Jersey Inst of Technology); Wen Zhang (New Jersey Institute of Technology); Qingquan Ma (New Jersey Inst of Technology); Neel Ahuja (Millburn High School)

Poster 55 - Copolymer blending for hydrophilization of gutter layers in thin film composite gas separation membranes
Arjun DeJesus (University of Connecticut)*; Noah Ferguson (University of Connecticut); Srivasta Bettahalli (University of Connecticut); Edward Wazer (University of Connecticut); Jeffrey R McCutcheon (University of Connecticut)

Poster 56 - Development of Machine Learning Models of Intermittent Operation of Multiple RO Systems based on Transfer AdaBoost and Adaptive Broad Learning Techniques
Yang Zhou (Shanghai University); Bilal M Khan (California State University San Bernardino); Nora Marki (University of California, Los Angeles); Yoram Cohen (Professor)*

Poster 57 - Performance of a low-pressure pilot-scale off-grid nanofiltration system
Katelynn M Carroll (University of Arizona)*

Poster 58 - Enhanced CO2 Separation Using Polymeric Membranes with Encapsulated Ionic Liquid
Eyad Qasem (King Abdullah University for Science and Technology)*; Suzana Nunes (King Abdullah University of Science and Technology (KAUST))

Poster 59 - Taxonomic Method for Characterization of Microbial Community in Full-Scale Reverse Osmosis Facilitates
Ruizhe Xu (the University of Texas at Austin)*

Poster 60 - Preparation of CO2-Selective Polydimethylsiloxane Mixed Matrix Membranes
Josef Schneider (University of Chemistry and Technology, Prague)*; Karel Friess (University of Chemistry and Technology, Prague)

Poster 61 - Molecular Engineering of Free volume in Membranes for Stationary Energy Storage
Toby Wong (Imperial College London)*

Poster 62 - An alternative application of 2D carbon-based membranes for efficient air separation and H2 purification from gas mixtures
Dániel Gardenö (UCT Prague)*; Karel Friess (UCT Prague); Zdenek Sofer (UCT Prague); Malte Kaspereit (FAU Erlangen-Nürnberg)

Poster 63 - Effect of pressure on the intrinsic permeability of ions in nanofiltration and reverse osmosis membranes
Mor Avidar (Technion)*; Liat Birnhack (Technion); Razi Epsztein (Technion)

Poster 64 - Exploring the energy of molecular partition into polyamide membranes using quartz crystal microbalance
oren ben porat (technion)*; Razi Epsztein (Technion); Liat Birnhack (Technion - Israel institute of technology)

Poster 65 - Thin Film Composite Ion Exchange Membranes by Electrospray
Yasmin Teixeira Bimbatti (UCONN)*; Deniz Ipekci (University of Connecticut); Xingyu Wang (University of Connecticut); Inara Oliveira (University of Virginia); Anna Harris (University of Virginia); Vian Yang (Georgia

Institute of Technology); Hazel Gerber (Georgia Institute of Technology); Baikun Li (University of Connecticut); Geoffrey M Geise (University of Virginia); Paul A Kohl (Georgia Institute of Technology); Jeffrey R McCutcheon (University of Connecticut)

Poster 66 - Catalytic nanoparticle functionalized commercial membranes for hydrogenation of PFAAs
Subrajit Bosu (University of Kentucky)*; Dibakar Bhattacharyya (U. Kentucky); Rollie Mills (U. Kentucky); Sam S Thompson (University of Kentucky); Angela Gutierrez (University of Kentucky)

Poster 67 - A Fractal Analysis Based Metrology to Quantify Porous Structure of Membranes
Riccardo Lovison (University of Colorado Boulder)*; Yifu Ding (Membrane Science, Engineering, and Technology (MAST) Center, Department of Mechanical Engineering, University of Colorado Boulder); Christina Carbrello (MilliporeSigma)

Poster 68 - Thin Film Composite Membranes from Polymers of Intrinsic Microporosity for CO2 Separation
Sunshine Iguodala (Imperial College London)*

Poster 69 - Exploring Deep Eutectic Solvent Resistance and Morphological Stability of Polyvinylidene fluoride (PVDF) and PVDF-based Blend Membranes for Sustainable Biomass Processing
Odianosen I Ewah (University of Kentucky)*; Isabel Escobar (University of Kentucky)

Poster 70 - Hydrophobic polyamide nanofilms provide rapid transport for crude oil separation
siyao li (Imperial College London); Ruijiao Dong (Shanghai Jiao Tong University); Valentina Elena Musteata (King Abdullah University of Science and Technology (KAUST)); Jihoon Kim (Korea Research Institute of Chemical Technology); Neel Rangnekar (ExxonMobil); JR Johnson (Exxon Mobil); Bennett D Marshall (ExxonMobil); Stefan Chisca (KAUST); Jia Xu (Ocean University of China); Scott Hoy (ExxonMobil); Benjamin McCool (ExxonMobil Research and Engineering); Suzana Nunes (King Abdullah University of Science and Technology (KAUST)); Zhiwei Jiang (Queen Mary University of London)*; Andrew Livingston (Queen Mary University of London)

Poster 71 - Supported liquid membrane contactor for efficient cesium extraction using crown either ionic liquid
Jintao He (Tiangong University)*; Jianxin Li (Tiangong University); Ranil Wickramasinghe (University of Arkansas)

Poster 72 - Direct Membrane Osmometry for Rapid Osmotic Pressure Measurements at High Molality
Mi Zhang (University of Connecticut)*; Jeffrey R McCutcheon (University of Connecticut); George Bollas (University of Connecticut); Edward Wazer (University of Connecticut); Wajeha Tauqir (University of Connecticut)

Poster 73 - Green Solvents for Upcycling Waste PVC into Ultrafiltration Membranes for Water Treatment
ATTA UR RAZZAQ (The University of Alabama)*; Milad Rabbani Esfahani (The University of Alabama)

Poster 74 - Polymer evolution and fabrication of PVDF membranes for contaminant removal from water
Rachel Kaplan (University of Kentucky)*; Minwoo Jung (Georgia Institute of Technology); Tequila Harris (Georgia Institute of Technology); Isabel Escobar (University of Kentucky)

Poster 75 - Protocol for purification of circular RNA from a self-splicing reaction by ultrafiltration
Karen d Guillen Cuevas (Clemson University); Marc Birtwistle (Clemson University); Scott Husson (Clemson University)*

Poster 76 - Polyamide Thin-film Composite Membranes for Gas Separations
Salman Ahmadipouya (University of Connecticut)*; Srivatsa Bettahalli (University of Connecticut); Jeffrey R McCutcheon (University of Connecticut)

Poster 77 - Real-Time Induced Magnetic Vibrations for Enhanced Fouling Mitigation
Jasneet K Pala (The University of Alabama)*; Milad Rabbani Esfahani (The University of Alabama); Ryan Tracy (The University of Alabama); Nima Mahmoodi (The University of Alabama)

Poster 78 - Visualization of Electrospray Technique for Membrane Manufacturing
Usama Javed J Sheikh (university of Connecticut)*; Jeffrey R McCutcheon (University of Connecticut); Srivatsa Bettahalli (University of Connecticut); Yuepeng

Zhang (Argonne National Laboratory); Feng Gao (Argonne National Laboratory)

Poster 79 - MORPHOLOGY AND CHEMISTRY CONTROL OF THIN FILM COMPOSITE MEMBRANES VIA ELECTROSPRAY 3D PRINTING

Mert Can Hacifazlioglu (University of Connecticut)*; Jeffrey R McCutcheon (University of Connecticut)

Poster 80 - Scaled Manufacturing of Supported Loose-NF Membranes with Eco-Friendly Solvents for Water Treatment Applications

David Lu (University of Kentucky)*; Kwangjun Jung (Georgia Institute of Technology); Ju Young Shim (Georgia Institute of Technology); Tequila Harris (Georgia Institute of Technology); Isabel Escobar (University of Kentucky)

Poster 81 - Patterned anion exchange membrane for improved performance of glucose fuel cells

Thomas A DosSantos DiSorbo (University of Colorado Denver | Anschutz)*

Poster 82 - Mixed binary alkali halide salt transport in PEGDA

Aubrey Quigley (The University of Texas at Austin)*; Everett S Zofchak (The University of Texas at Austin); Kate Dycha (The University of Texas at Austin); Nathaniel Lynd (The University of Texas at Austin); Benny Freeman (The University of Texas at Austin)

Poster 83 - Influence of pH, Ionic Strength, and Operational Parameters on Capture Mechanisms in Nanoparticle Filtration

Lauren Tice (MilliporeSigma)*; Joseph Hersey (MilliporeSigma); Surya Karla (Rensselaer Polytechnic Institute); Mirco Sorci (Rensselaer Polytechnic Institute); Joel Plawsky (Rensselaer Polytechnic Institute); Sal Giglia (MilliporeSigma); Georges Belfort (Rensselaer Polytechnic Institute)

Poster 84 - Catalytic Membrane Reactor Enabled by H2-Selective Carbon Molecular Sieve Membrane for Blue Hydrogen Production

Kai Chen (University at Buffalo)*; LEIQING HU (University at Buffalo); Gengyi Zhang (The State University of New York at Buffalo); Haiqing Lin (The State University of New York at Buffalo)

Poster 85 - Ion-ion Separations in Biomimetic Water Channels

Tyler J Duncan (The University of Texas at Austin)*; Harekrushna Behera (The University of Texas at Austin); Nico Marioni (The University of Texas at Austin); Meron Tadesse (University of Texas Austin); Paul R Irving (University of Texas at Austin); Everett S Zofchak (The University of Texas at Austin); Manish Kumar (University of Texas at Austin); Venkat Ganesan (University of Texas Austin)

Poster 86 - Incorporation of Gutter Layers in Reverse Osmosis and Nanofiltration Thin Film Composite Membranes

Martina Jagielski (University of Connecticut)*; Mert Can Hacifazlioglu (University of Connecticut); Jeffrey R McCutcheon (University of Connecticut)

Poster 87 - Structure-property relationships in intercalated graphene oxide nanofiltration membranes for enhanced separation of salts and organics from aqueous streams

Muskan Sonker (Georgia Institute of Technology)*; Sankar Nair (Georgia Institute of Technology); Kaung Su Khin Zaw (Georgia Institute of Technology); Hazel P. Dhruve (, Georgia Institute of Technology); Mahsa Abbaszadeh (Albemarle); Madeline P. Garell (Georgia Institute of Technology); Michael B. Salerno (Phillips 66 Research Center); Marta Hatzell (Georgia Institute of Technology); Meisha L. Shofner (Georgia Institute of Technology)

Poster 88 - Tuning ionic selectivity in nanoporous graphene membranes using atomic layer deposition and polyelectrolyte assembly

Simar K Mattewal (MIT)*; Rohit Karnik (MIT)

Poster 89 - Strengthening Isoporous Polymer Membranes through Chain Entanglements

Adam N Mann (University of Texas at Austin)*; Benny Freeman (University of Texas at Austin); Gabriel Sanoja (University of Texas at Austin)

Poster 90 - Computer vision-based measurement of pure- and mixed-gas swelling in polymer membranes exhibiting configurational free volume

William Box (University of Oklahoma)*; Lucas Condes (University of Oklahoma); Matthew T Webb (University of Oklahoma); Agboola OS Suleiman (University of Notre Dame); Michele Galizia (University of Oklahoma); Ruilan Guo (University of Notre Dame)

Poster 91 - Impact of feedwater composition on the removal of short, ultrashort, and neutral PFAS in nanofiltration

Yukai Tomsovic (University of Colorado, Boulder)*; Anthony Straub (University of Colorado Boulder); Kyle

Doudrick (University of Notre Dame)

Poster 92 - Sulfonated Polyamide Thin Film Composite Membranes for Electrodialysis
Deniz Ipekci (University of Connecticut)*; Yasmin Teixeira Bimbatti (UCONN); Xingyu Wang (University of Connecticut); Inara Oliveira (University of Virginia); Anna Harris (University of Virginia); Geoffrey M Geise (University of Virginia); Baikun Li (University of Connecticut); Jeffrey R McCutcheon (University of Connecticut)

Poster 93 - Maximizing the Wetting Resistance of Fluorine-free Omniphobic Membranes for Hypersaline Wastewater Desalination
Yinchuan Yang (University of British Columbia)*; Sifat Kalam (University of British Columbia); Sadaf Shabanian (University of British Columbia); Kevin Golovin (University of Toronto); Yalei Zhang (Tongji University); Jongho Lee (University of British Columbia)

Poster 94 - Designing the fluorinated superfine-activated carbon adsorptive membrane for selective removal of long- and short-chain PFAS
Medha Kasula (The University of Alabama)*; Milad Rabbani Esfahani (The University of Alabama)

Poster 95 - Gas Permeation and Separation Characteristics of Microporous TpHz COF Membranes Synthesized by Substrate-Assisted Interfacial Polymerization
Jose E Lopez-Cazares (Arizona State University)*; Jerry Lin (Arizona State University); Kailong Jin (Arizona State University); Kyungtae Kim (Los Alamos National Lab)

Poster 96 - Greener TEAMs: Tethered Electrolyte Active-Layer Membranes Produced by Surface-Initiated Free Radical Polymerization
Elnur Jabiyev (Auburn University)*; Mohammad Mehdi Pour (Auburn University); Cassandra J Porter (Auburn University)

Poster 97 - IgG-BSA Separation and Purification by Internally Staged Ultrafiltration
Lixin Feng (New Jersey Institute of Technology); Yufeng Song (New Jersey Institute of Technology); Sagnik Basuray (New Jersey Institute of Technology); Kamalesh K Sirkar (New Jersey Institute of Technology)*

Poster 98 - Performance and Environmental Evaluation of Biochar-based Membranes Synthesized from Various Petroleum- and Bio-derived Solvents
Abelline K Fionah (University of Kentucky)*; Isaac Oluk (University of Kentucky); Isabel Escobar (University of Kentucky)

Poster 99 - Thin Film Nanocomposite Membranes Embedded with ZIF-67 for Desalination and Boron Removal
Suer Kurklu Kocaoglu (The University of Texas at Austin)*; Benny Freeman (The University of Texas at Austin); S. Birgül Tantekin-Ersolmaz (Istanbul Technical University)

Poster 100 - Elucidating the Structure-Property-Performance Relationship of Incorporating Waste Polymethyl Methacrylate in Polyvinylidene Fluoride Membrane Fabrication
Joseph T Danner (The University of Alabama)*; Liam Papier (The University of Alabama); Steven T Weinman (The University of Alabama)

Poster 101 - Long-term Operation of Pilot-scale Membrane Bioreactor – Membrane Distillation for Potable Water Reuse
Luke Presson (University of Arizona)*; Michael Hegetschweiler (University of Arizona); Andrea Achilli (University of Arizona)

Poster 102 - In-Situ Synthesis of ZIF-L MOF on UF Membranes for Efficient Microplastics and Dyes Removal
Himangshu Mondal (The University of Alabama)*; Medha Kasula (The University of Alabama); Milad Rabbani Esfahani (The University of Alabama)

Poster 103 - Unlocking the power of a gutter layer: A new frontier in dye-desalination using amine-crosslinked nanoporous graphene oxide (aHGO) platform.
Ameya Tandel (State University of New York, Buffalo, NY)*; Haiqing Lin (The State University of New York at Buffalo)

Poster 104 - Development of a high-performance 30nm NIPS PTFE membrane
Karl Ashkar (Pall Corporation)*; SIAVASH DARVISHMANESH (Pall corporation); Steven Gardner

(Pall Corporation)

Poster 105 - Designing membranes using bottlebrush poly(1,3-dioxolane) acetate for CO2 capture
Fathy Attia (The State University of New York at Buffalo)*; Thien N Tran (University at Buffalo, The State University of New York); Erda Deng (State University of New York at Buffalo); Bhanuprakash Valluri (University at Buffalo); Liang Huang (The State University of New York at Buffalo); Haiqing Lin (The State University of New York at Buffalo)

Poster 106 - XpressKleen 1 nm expanded PTFE membrane for harsh chemical filtration
Brenner Zhefei Li (Pall Corporation)*; SIAVASH DARVISHMANESH (Pall corporation); Catherine Agcaoili (Pall Corporation)

Poster 107 - Enabling high-selectivity carbon molecular sieve hollow fiber membranes via precursor selective layer engineering
Brian Leonard (West Virginia University)*; Oishi Sanyal (West Virginia University)

Poster 108 - Lithium Selective Artificial Ion Channels for Lithium Separation
Harekrushna Behera (The University of Texas at Austin)*; Tyler J Duncan (The University of Texas at Austin); Laxmicharan Samineni (University of Texas at Austin); Hyeonji Oh (University of Texas at Austin); Venkat Ganesan (The University of Texas at Austin); Manish Kumar (University of Texas at Austin)

Poster 109 - Using high-throughput membrane characterization to elucidate how nuanced changes in membrane chemistry affects solute transport
Laurianne Lair (University of Notre Dame)*

Poster 110 - Real-Time, Non-Destructive Monitoring of Supported Liquid Membranes for Biochemical Recovery Using Electrochemical Impedance Spectroscopy
Hiroki Fukuda (University of British Columbia); Jongho Lee (University of British Columbia)*

Poster 111 - Delamination Mitigation of Modified Bilayer Membranes using Single Slot Die Roll-to-Roll Manufacturing Process
Minwoo Jung (Georgia Institute of Technology)*; Rachel Kaplan (University of Kentucky); Isabel Escobar (University of Kentucky); Tequila Harris (Georgia Institute of Technology)

Poster 112 - Modeling Hybrid Systems for Industrial Carbon Capture
Carol Jerotich (University of Pittsburgh)*; Katherine Hornbostel (University of Pittsburgh)

Poster 113 - Low-cost and scalable automated flat-sheet membrane casting: an open-source, advanced manufacturing approach
Charles-François de Lannoy (McMaster)*; Nathan Mullins (McMaster University); David Latulippe (McMaster University)

Poster 114 - Separation of siloxane from silicone fluid mixture using polymeric membrane
Hammed A Balogun (Georgia Institute of Technology)*; Reza Panah (Dow Chemical Company, Midland); Dimitris Katsouli (Dow Chemical Company, Midland); Jay Rose (Dow Chemical Company, Midland); William Koros (Georgia Institute of Technology); Ryan Lively (Georgia Tech)

Poster 115 - Evidence for solution-diffusion in pressure-induced transport of solvent through dense polymer membranes
Kevin Reimund (University of Texas at Austin)*; Rahul Sujanani (The University of Texas at Austin); Jorge Hernandez (The University of Texas at Austin); Kristofer Gleason (The University of Texas at Austin); Manish Kumar (University of Texas at Austin); Benny Freeman (The University of Texas at Austin)

Poster 116 - CFD modeling of high-flux plate-and-frame membrane modules for industrial carbon capture
Cheick Dosso (Carnegie Mellon University)*; Hector A Pedrozo (Carnegie Mellon University); Thien Tran (National Energy Technology Laboratory); Lingxiang Zhu (National Energy Technology Laboratory); Victor A Kusuma (NETL); David Hopkinson (NETL); Lorenz Biegler (Carnegie Mellon University); Grigorios Panagakos (Carnegie Mellon University)

Poster 117 - Optimizing Carbon Capture Plants with Hollow Fiber Membrane Technology: Integration of CFD Rigorous Models through Trust-Region Methods

Hector A Pedrozo (Carnegie Mellon University); Lorenz Biegler (Carnegie Mellon University); Grigorios Panagakos (Carnegie Mellon University)*

Poster 118 - Demonstration of a Real-Time Maximum Power Point Tracker for Salt Gradient Osmotic Power Systems
Sanjana Chintalacheruvu (Oakland University); saber khanmohammadi (Oakland University)*; Jonathan Maisonneuve (Oakland University)

Poster 119 - Removal of PFOS in the Presence of Humic Acid Using Activated Carbon-Functionalized Adsorptive Thin-film Nanocomposite Membranes
Isabella Le (The University of Alabama)*; Medha Kasula (The University of Alabama); Milad Rabbani Esfahani (The University of Alabama)

Poster 120 - Enhancing the performance of green solvent synthesized ABS membranes using co-solvents
Sulaiman Dhameri (University of Kentucky)*; Emily Ingram (University of Kentucky); Anastasiia Aronova (University of Kentucky); Endras Fadhilah (University of Kentucky); Gosia Chwatko (University of Kentucky)

Poster 121 - A Membrane Contactor Enabling Energy-Efficient CO2 Capture from Point Sources with Physical Solvent
Nicholas Gregorich (Oak Ridge National Laboratory)*; Mary Irwin (Oak Ridge National Laboratory); Zachary Coin (Oak Ridge National Laboratory); Stephen JA DeWitt (Arkema Inc); Ramesh Bhave (Ornl); David Sholl (Oak Ridge National Laboratory); Syed Islam (Oak Ridge National Laboratory)

Poster 122 - Zr-based MOF-NF-Membrane for Selective Removal of Selenium from Produced water
Sweta A Modak (The University of Alabama)*; Milad Rabbani Esfahani (The University of Alabama)

Poster 123 - Anti- swelling Characteristics Enhanced via Self-Cross-Linking in MXene-Interlaced Graphene Oxide Membranes
Saurabh Kr Tiwary (University Of Houston)*; Alamgir Karim (University of Houston)

Poster 124 - Membrane-Based CO2 Removal From Desalination Plants
Marina Hildebrandt (University of Pittsburgh)*; Katherine Hornbostel (University of Pittsburgh)

Poster 125 - Advanced functionalized polyether ether ketone/polysulfone membranes with 2D nanomaterials for enhanced water treatment performance
Muhammad Usman Yousaf (University of Kentucky)*; Nerissa Kiprotich (University of Kentucky); Anika Azme (University of Nebraska, Lincoln); Lucca Madeo Cortarelli (University of Kentucky); Nirupam Aich (University of Nebraska, Lincoln); Olga Tsyusko (University of Kentucky); Isabel Escobar (University of Kentucky)

Poster 126 - Mechanisms of Flux Decline in Crossflow Ultrafiltration of Whey Protein: Comparing Different Models to Experimental Data
Burcu Ozdemir (UCLA)*; Eric Hoek (UCLA)

Poster 127 - Electrifying membrane distillations: Localized Joule heating of laser-induced graphene membranes
Md Hasib Al Mahbub (Texas Tech University)*; Mahdi Malmali (Texas Tech University)

Poster 128 - From surface wetting to internal wetting: understanding wetting front progression in membrane distillation
Shounak G Joshi (University of Southern California)*; Amy Childress (University of Southern California); Allyson McGaughey (University of New Mexico)

Poster 129 - Catalytic Membranes Based on rGO for Micropollutant Treatment Enabled by Peracetic Acid
Erda Deng (State University of New York at Buffalo)*; Haiqing Lin (The State University of New York at Buffalo)

Poster 130 - Understanding the dopant effects on the selective transport of cations in conductive polymers
Meng Shen (California State University, Fullerton)*

Poster 131 - Elucidation of the Effect of Hydrophobic/Hydrophilic Domains in Polyamide Thin-Film Composite Reverse Osmosis Membranes
Mohammad Mehdi Pour (Auburn University)*; Elnur

Jabiyev (Auburn University); Cassandra J Porter (Auburn University)

Poster 132 - Chemistry-structure-performance relationships of polyamide membranes for lithium recovery
Rebecca Lee (University of Texas at Austin)*; Manish Kumar (University of Texas at Austin); Abhishek Roy (NREL); Nurshaun Sreedhar (University of Texas at Austin); Mou Paul (NREL)

Poster 133 - Membrane heterogeneity alters our interpretation of effective energy barriers to transport
Nathanael S Schwindt (University of Colorado Boulder)*; Mor Avidar (Technion); Razi Epsztein (Technion - Israel Institute of Technology); Anthony Straub (University of Colorado Boulder); Michael Shirts (University of Colorado Boulder)

Poster 134 - Freezing desalination using serpentine pipe: How freezing time and washing technique improve salt rejection and production ratea
Siriwat Singto (U. of Hawai'i); Woochul Lee (U. of Hawai'i); Albert Kim (U. of Hawai'i)*

Poster 135 - Assessing stability and toxicity of the polymeric membranes fabricated with petroleum- and bio-derived solvents.
Lucca Madeo Cortarelli (University of Kentucky)*; Muhammad Usman Yousaf (University of Kentucky); David Lu (University of Kentucky); Isabel Escobar (University of Kentucky); Anika Azme (University of Nebraska-Lincoln); Nirupam Aich (University of Nebraska-Lincoln); Olga Tsyusko (University of Kentucky)

Poster 136 - Adjustable microstructure design of a polyphthalonitrile precursor and the derived novel carbon molecular sieve membrane for gas separation
Chun-Po Hu (Nanyang Technological University)*

Poster 137 - Enhancing Electrodialysis with Thiol-ene Polymer Network-Based Anion-Exchange Membranes
Sara Abu-Obaid (University of Toronto)*; Jocelyn Riet (University of Toronto); Jay Werber (University of Toronto)

Poster 138 - Fabrication of Polysulfone/Sulfonated-polyether-ether-ketone Membranes for Removal of Organic Contaminants from Water
Nerissa Kiprotich (University of Kentucky)*; Muhammad Usman Yousaf (University of Kentucky); Isabel Escobar (University of Kentucky)

Poster 139 - Computational insights from the polymer/MOF interface in Mixed-Matrix Membranes
Moein Hajian Zaker Damghan (Texas Tech University)*; Hyunhee Lee (MIT); Zachary Smith (MIT); Joshua Howe (Texas Tech University)

Poster 140 - Experimental and Modeling Results for the Rejection of Ultrashort Per-and Polyfluoroalkyl Substances
Aron Griffin (Colorado School of Mines)*; Timothy Strathmann (Colorado School of Mines); Christopher Bellona (Colorado School of Mines)

Poster 141 - Compaction of high-pressure water treatment membranes: real-time quantification and support layer contribution
Weijian Ding (University of Southern California)*; Kexin Ma (University of Southern California); Amy Childress (University of Southern California)

Poster 142 - Simplified Synthesis Methods for Membrane-Protein Nanosheets and Biomimetic Membranes
Ronald J Vogler (The University of Texas at Austin)*; Katherine Kimball (The University of Texas at Austin); Dominic Bujanos (The University of Texas at Austin); Raman Dhiman (The University of Texas at Austin); Benny Freeman (The University of Texas at Austin); Manish Kumar (University of Texas at Austin)

Poster 143 - Electrospun Fibrous Membranes for Bioreactor Harvesting of Viral Vectors
Nimisha Tandon (University of Arkansas); Ranil Wickramasinghe (University of Arkansas); Xianghong Qian (University of Arkansas)*

Poster 144 - The Effects of Pressure and Process Interruption on The Location of Virus Particles Captured during Virus Filtration
Wenbo Xu (University of Arkansas)*; Hironobu Shirataki (Asahi-Kasei); Ranil Wickramasinghe

(University of Arkansas); Xianghong Qian (University of Arkansas)

Poster 145 - Enhancing Water Evaporation in Membrane Processes Using Hydrogel Coatings

Omar Laris (CU Boulder)*

Poster 146 - The Effect of Crosslinker Concentration on Competitive Release Kinetics and Physical Characteristics of Thermo-Responsive, Lignin-Based Soft Composites

Missoury Wolff (Clemson University)*; Eric Davis (Clemson University)

Poster 147 - Hydrogen- and acetylene-stable silver ion facilitated transport membranes: the effect of salt content on fractional free volume and gas permeability

Matthew N Davenport (The University of Texas at Austin)*; Tangqiumei Song (UT Austin); Justin J Rosenthal (The University of Texas at Austin); Cara Doherty (CSIRO); Joan Brennecke (The University of Texas at Austin); Benny Freeman (The University of Texas at Austin)

Poster 148 - Insights into methods of determining selectivity of porous ultrafiltration membranes

Suzana Ivandic (UT Austin)*; Mostafa Nassr (UT Austin); Alishba Amjad (UT Austin); Ben Pedretti (MIT); Nathaniel Lynd (The University of Texas at Austin); Lynn Katz (The University of Texas at Austin); Benny Freeman (The University of Texas at Austin)

Poster 149 - Ultrafiltration Membranes from Polymerized Lyotropic Liquid Crystals for Bioseparation Applications

Seyed Mostafa Tabatabaei (University of Oklahoma); Reza Foudazi (University of Oklahoma)*

Poster 150 - Targeted Separation Scheme of Polyurethane Depolymerization Products

Taysha B Telenar (Arizona State University)*; Timothy Long (Arizona State University); Matthew Green (Arizona State University)

Poster 151 - Towards the Synthesis of Mixed Matrix Membranes Using Vapor Phase Infiltration

Seancarlos Gonzalez (University of Washington)*; Sarah West (University of Washington); Yuri Choe (University of Washington); Kellen McKinney (University of Washington); Cecilia Osburn (University of Washington); David S Bergsman (University of Washington)

Poster 152 - Role of salt additives in tuning the structure of β-cyclodextrin-based membranes

Jinting Hu (KU Leuven)*; Sem Bleus (KU Leuven); Linde Achten (KU Leuven); Wim Dehaen (KU Leuven); Xing Yang (KU Leuven)

Poster 153 - Fabrication of asymmetric nanofiltration membranes prepared from sulfonated poly(arylene ether sulfone) for monovalent-divalent ion separation

Nurshaun Sreedhar (University of Texas at Austin)*; Beatrice Tremblay (University of Virginia); Rebecca Lee (University of Texas at Austin); Karli Gaffrey (Colorado School of Mines); Geoffrey M Geise (University of Virginia); Mou Paul (National Renewable Energy Laboratory); Manish Kumar (University of Texas at Austin); Abhishek Roy (NREL)

Poster 154 - NEW CLASS OF IMIDAZOLIUM-BASED IONENES FOR GAS SEPARATION MEMBRANES

Sudhir Ravula (University of Alabama)*; Jason Bara (University of Alabama)

Abstracts – Oral Presentations, Monday Morning

Presenting authors are indicated by an asterisk*.

Electrochemical Separations

8:30 AM
Oral 1 - Understanding Polyfluoroalkyl Substance Selectivity in Anion Exchange Membrane Systems

Sam S Thompson (University of Kentucky)*; Angela Gutierrez (University of Kentucky); Rollie Mills (U. Kentucky); Dibakar Bhattacharyya (U. Kentucky)

The widespread production and use of multi-fluorinated carbon-based substances for a variety of purposes has contributed to the contamination of global water supply in recent decades. Per- and polyfluorinated alkyl substances (PFAS) are an anthropogenic class of chemicals imbued with desirable industrial properties that consequentially result in their difficult remediation once exposed to the environment. Common remediation strategies incorporate adsorption and/or ion exchange technologies to separate such contaminants from water, but often suffer from limitations due to low mass transfer efficiencies, among other issues. Membrane systems use many of the same separation methods currently in use today while enhancing operational potential of adsorption or ion exchange functionalities by boosting mass diffusion through convective flow. Furthermore, these improvements are applicable to the regeneration of such membrane systems, which is a costly drawback to conventional water purification techniques. To this end, commercial microfiltration and ultrafiltration membranes were modified using pore functionalization to incorporate an anion exchange moiety within the membrane matrix. This functionalization was performed with multiple amine-containing polymer networks ranging from weak to strong basic residues. Modified membranes maintained water permeance above 40 LMH/bar. Separation of PFAS from aqueous streams was analyzed for both "long-chain" and "short-chain" analytes, perfluorooctanoic acid and perfluorobutyric acid, respectively. Synthesized membranes demonstrated as high as 90% first-pass rejection of perfluorooctanoic acid. Ion exchange affinity of competitive solution-phase ions was also analyzed. This research advances a versatile membrane platform for environmentally relevant applications that seek to help the global availability of safe drinking water. Research funding was provided by the NIEHS-sponsored University of Kentucky Superfund Research Center.

8:50 AM
Oral 2 - Nanofiltration membranes with thermo- and pH-responsive behavior from lyotropic liquid crystal templating

Younes Saadat (University of Oklahoma); Kyungtae Kim (Los Alamos National Laboratory); Reza Foudazi (University of Oklahoma)*

In this work, we use normal hexagonal phase (H1) of lyotropic liquid crystals (LLCs) as a template to produce orderly packed nanofibrous membranes. Compared to conventional methods like non-solvent induced phase separation (NIPS), LLC templating approach for making membranes is a more sustainable approach. The obtained membrane has a continuous 3D transport pathway and can alter its nanofiltration properties in response to changes in temperature and pH. The formulation includes Pluronic P84-diacrylate (P84DA), a thermoresponsive component that acts as both macromer and structure-directing amphiphile. P84DA facilitates changes in membrane pore size with temperature due to its thermo-responsiveness when is in contact with water. Furthermore, the precursor contains acrylic acid (AAc) as the charged component, which upon copolymerization with P84DA, not only enables ion separation through Donnan exclusion, but also imparts pH-responsive behavior for the separation of ionic species. We show that the nanofiltration membrane has separation properties adjustable with temperature and pH with exceptional resistance to fouling by various solutes due to its highly hydrophilic surface. Furthermore, the membrane shows an outstanding sulfate over chloride ion selectivity, which is a requirement for salt fractionation applications.

9:10 AM

Oral 3 - Charge-patterned nanofiltration membranes with immobilized polystyrene sulfonate particles and polyethylene imine

Ali Behboudi (The Pennsylvania State University)*; Philipp Jahn (Lehrstuhl für Technische Chemie II, Universität Duisburg-Essen, 45141 Essen, Germany); Mathias Ulbricht (Lehrstuhl für Technische Chemie II, Universität Duisburg-Essen, 45141 Essen, Germany)

Removing salinity has always been a challenge for wastewater treatment applications. Utilizing nanofiltration (NF) membranes is a promising approach to overcome this challenge. However, currently available NF membranes are less effective in monovalent salt removal. In this study, work toward the initial aim for fabricating charge mosaic membranes led to charge-patterned NF membranes prepared on polyether sulfone (PES) and polyacrylonitrile (PAN) support membranes with interestingly similar performance in both mono- and multivalent salts rejection. The membranes were fabricated by double layer assembly of negatively charged polystyrene sulfonate nanoparticles (PSS) immobilized in a polyvinyl alcohol (PVA) layer, followed by an additional positively charged polyethylene imine (PEI) layer on top. Both PVA and PEI layers were crosslinked using glutaraldehyde (GA) in a final step. The support membrane along with nanoparticle, PVA and PEI concentrations, as well as feed pH and salt concentration play a significant role for the separation performance of these types of membranes. The results indicate that the proposed charge-patterned NF membranes fabricated using 0.5 wt.% PVA, 0.05 wt.% PSS as the first assembled layer followed by a sealing layer assembled via 0.5 wt.% PEI solution was able to deal with monovalent salt (NaCl) even better than that of the multivalent salts (Na2SO4, MgSO4, and MgCl2), with the salt rejection of 80% and 75%, respectively, at a permeance of 3.1 LMH/bar on the PES substrate and 5.5 LMH/bar in the case of PAN substrate.

9:30 AM

Oral 4 - Treating Acidic Metal Wastewaters with Ceramic Ion Exchange Membranes

Emily Rabe (Membrion)*; Rachel Malone (Membrion); Danny Bondurant (Membrion); Stephanie Candelaria (Membrion); Olivia Lenz (Membrion); Gregory M Newbloom (Membrion, Inc.)

Electrodialysis (ED) is a well-developed technology for ion removal using ion exchange membranes (IEMs) and an applied electric field. However, many industries generate harsh wastewater streams that are incompatible with traditional organic polymer-based ion exchange membranes. Membrion® has developed novel ceramic-based IEMs offering several advantages over traditional polymer membranes: high performance in low pH, chemical resistance to oxidizers, and a rigid structure which minimizes swelling. Using our patented silane-based sol-gel process we produce both anion exchange (AEM) and cation exchange membranes (CEM) via roll-to-roll processing at ambient temperatures. We have benchmarked our novel IEM membrane performance against traditional IEMs via permselectivity, area-specific resistance, and ED stack testing.

In the context of industrial wastewater treatment, ED has demonstrated the ability to reduce the concentration of salts, metals, and minerals by up to 99.5% at a water recovery of 65 – 98%, where the desalinated effluent is suitable for discharge or reuse. We have successfully piloted our technology at a semiconductor foundry treating low pH copper wastewater to meet permitted discharge requirements. As shown in Figure 1, we maintained stable operation for over 50 batches of waste. The TDS variability, low-pH, oxidizers, chelators and particles significantly limited alternative treatment options, leading to off-site trucking and disposal of the wastewater which often involved frequent multi-hundred-mile trips. Leveraging this technology has the potential to significantly reduce a manufacturer's carbon footprint by mitigating trucking wastewater for offsite treatment, and instead provides the opportunity for water reuse and increased circularity.

9:50 AM

Oral 5 - Synthesis and Testing of Electrochemical Activity in a Polyether Sulfone-Based Ultrafiltration Membrane for Contaminant Removal

Utkarsh Misra (IIT Bombay)*; Kritika Jashrapuria (Indian Institute of Technology Bombay); Swatantra Pratap Singh (IIT BOMBAY)

With the advent of modern pollutants in freshwater sources and increased demand, treating wastewater remains a viable option for reuse. Membrane technology has been a flag bearer for treating untreatable pollutants with advanced oxidation processes. However, the membranes face a severe issue of fouling and concentrating stream, limiting the process utilization. The hybrid of advanced oxidation with filtration can solve these problems sustainably. In this paper, an electrochemically active membrane was fabricated using laser-induced graphene(LIG) and polyether-sulfone(PES) polymer. The fabricated membrane provides the sites for degrading and transforming the pollutants. Despite ultrafiltration, these membranes have shown removal of Cr(VI) in the nanofiltration range on applying small electrical potential. The membranes are also fouling resistant due to the properties of LIG in the matrix, showing flux recovery ratios of ~90%. The fabricated membranes were tested using potentiostat for the electrochemical activity and crossflow filtration mode for the Cr(VI) pollutant removal. The membrane shows the removal of ~94% for Cr(VI) at a higher water permeability of ~150 LMH bar-1, with around 67% less biomass accumulation than pristine membranes. LIG provides a maximum surface roughness of ~ 100 nm in the membranes, which could be responsible for the antifouling properties. Further, the membranes were characterized using AFM, FTIR, SEM, and Contact angle. In conclusion, the fabricated membrane breaks the permeability selectivity trade-off using a fouling resistance membrane in a facile and scalable manner.

10:10 AM

Oral 6 - Electroconductive Filters and Membranes for Desalination and Wastewater Treatment

Swatantra Pratap Singh (IIT BOMBAY)*

Biofouling and selectivity permeability trade-offs are the leading challenges in membrane-based desalination and wastewater recycling. Electro-conductive membranes have sown its potential for controlling biofouling and selectivity permeability trade-offs. Laser-induced graphene (LIG) is a single-step, chemical-free method to fabricate graphene surfaces with a 10.6 um CO¬2 laser. Single-step conversion of polyethersulfone (PES) membranes to LIG filters and further surface modification of these filters leads to UF composite membranes that have been demonstrated to have excellent catalytic and anti-biofouling properties. These electrically conductive filters and membranes achieved 6-log removal of mix culture of bacteria with an applied voltage; additionally, the antibiotic rejection (~90%) and removal (~85%) were achieved and further improved with radical generation in the presence of metal ions. Furthermore, these surfaces have been shown to have anti-biofouling properties due to their texture, and the biofouling has been characterized by Confocal Laser Scanning Microscopy (CLSM). In the resister mode, their performance as the interfacial evaporator with multiple stacks of LIG filters was excellent. These single-step laser-printed filters and membranes have great potential in desalination and wastewater treatment.

10:30 AM

Oral 7 - Performance of Commercial Bipolar Membranes for Ocean Alkalinity Enhancement Application

Bassel A. Abdelkader (McMaster University)*; Charles-François de Lannoy (McMaster University)

Decarbonization of our environment requires reducing emissions as well as the rapid scale-up of carbon dioxide removal technologies (CDR). Ocean Alkalinity Enhancement (OAE) is one of the most promising techniques for CDR, but it requires significant quantities of dilute alkaline substance to be produced at low cost and from decarbonized energy. Conventional methods rely on either mining and transportation, which in many cases may negate the carbon removal benefits, or chlor-alkali processes to produce high-concentration of sodium hydroxide, which are energy intensive. Bipolar membrane electrodialysis

(BMED) theoretically offers a more energy- and cost-efficient approach for generating NaOH. A techno-economic analysis identified the membrane operation as one of the major cost hurdles to BMED OPEX. Despite the variety of bipolar membranes (BPMs) available in the market, there is a lack of comprehensive characterization of their properties. This study systematically characterizes several commercial bipolar membranes using different configurations, examining membrane properties, and overall performance within BMED systems. Two process configurations, closed loop, and feed and bleed, are evaluated. Preliminary findings reveal the significant impact of high feed concentration on membrane performance, indicating an upper bound of practical feed concentrations. For certain membranes tested, the NaOH production deteriorated, in some cases by up to two-thirds after three weeks of operation due to severe scaling. Results from this research will improve understanding of BMED systems, guiding future research strategies for successful OAE implementation. This systematic characterization of commercial bipolar membranes is essential for enhancing the efficiency and viability of OAE processes, contributing to sustainable ocean alkalinity enhancement efforts.

Advanced Membranes for Gas and Vapor Separations I

8:30 AM

Oral 8 - Enhanced H2 transport in polybenzimidazole by blending with a highly permeable polyimide

Narjes Esmaeili (The State University of New York at Buffalo)*; Leiqing Hu (University at Buffalo); Erda Deng (State University of New York at Buffalo); Haiqing Lin (The State University of New York at Buffalo)

Polybenzimidazole (PBI) with H2 permeability of 2.0 Barrer and H2/CO2 selectivity of 10 at 35 oC is a leading membrane material for H2/CO2 separation. However, the H2 permeability is too low for practical applications. Herein we demonstrate that PBI can be blended using a polyimide (6FDA-DAM) with H2 permeability of 610 Barrer and H2/CO2 selectivity of 1.1 at 35 oC, leading to a macrophase-separated structure containing discontinuous 6FDA-DAM phase and significantly improved H2 permeability without a loss of H2/CO2 selectivity. For instance, introducing 10-40 mass% 6FDA-DAM in PBI results in a discontinuous 6FDA-DAM phase, as confirmed by the SEM images, the two d-spacings from the WAXD patterns, and two glass transition temperatures from the DSC curves. Adding 10 mass% 6FDA-DAM in PBI increases H2 permeability by 40% from 2.0 to 2.8 Barrer while retaining H2/CO2 selectivity at 13 at 35 oC, and adding 40 mass% 6FDA-DAM increases H2 permeability to 7.4 Barrer by 270% without the loss of H2/CO2 selectivity. These results can be satisfactorily described using the Maxwell model. The effect of the compositions and temperature on gas sorption, diffusion, and permeation is thoroughly investigated. The blend containing 40 mass% 6FDA-DAM exhibits H2 permeability of 120 Barrer (four times that for pure PBI) and H2/CO2 selectivity of 10 at 150 oC, surpassing Robeson's upper bound. We will also discuss the long-term stability test of the sample with simulated syngas mixtures containing water vapor. This work demonstrates that polymer blending with macrophase separation can effectively improve gas permeability and overcome the permeability/selectivity tradeoff.

8:50 AM

Oral 9 - Tightening ultramicropores of structurally engineered carbon molecular sieve membranes for precise gas separation

Zhongyun Liu (Georgia Institute of technology)*; William Koros (Georgia Institute of Technology)

Advanced carbon molecular sieve (CMS) membranes with ultraselectivity are key needs for industrial gas separations. In this study, we developed CMS membranes with fine-tuned ultramicropores by strategically engineering structural features, including bulk groups and weak links in the backbone of polyimide precursors. Specifically, CMS membranes were fabricated using 6FDA-DAM, 6FDA-DAPI-5&6, and Matrimid as precursors with a pyrolysis temperature of 900 °C. Our findings demonstrate that intentional control over bulk groups and weak bonds in the precursor backbone allows for the precise tailoring of ultramicropores in the structurally engineered CMS membranes. Such engineered ultraselective ultramicropores enable simultaneously enhanced diffusion selectivity and sorption selectivity, making the developed CMS membranes highly attractive for various critical gas separations, particularly for their applications in hydrogen extraction modules (HEMs) and the recovery of dilute helium from natural gas feeds. Furthermore, this presentation will delve into how the engineered ultraselective ultramicropores in CMS membranes influence the contribution of entropic selectivity and energetic selectivity to the increased diffusion selectivity. Our study provides fundamental insights into the development of ultraselective CMS membranes tailored for precise gas separations.

9:10 AM

Oral 10 - Computer vision-based measurement of pure- and mixed-gas swelling in polymer membranes exhibiting configurational free volume

William Box* (University of Oklahoma); Lucas Condes (University of Oklahoma); Matthew T Webb (University of Oklahoma); Agboola OS Suleiman (University of Notre Dame); Michele Galizia (University of Oklahoma); Ruilan Guo (University of Notre Dame)

Although swelling data is needed to assess the long-term stability of polymer membranes, the swelling propensity of next-generation polymer membranes exhibiting configurational free volume has yet to be addressed thus far. To fill this literature gap, we investigate pure and mixed-gas swelling of a series of thermally rearranged polybenzoxazoles (TPBOs) and their poly(o-hydroxyimide) precursors exhibiting systematically varied amount of configurational free volume. To accomplish this, we developed a brand new experimental dilatometry approach utilizing computer vision techniques, which is capable of correcting most systematic limitations of traditional methods used to measure swelling and achieve superior measurement accuracy. The proposed design also allows for the measurement of multiple samples at once and the correction of multiple sources of systematic error, thereby improving the accuracy of comparison between measurements. The newly proposed approach helps shed fundamental light on the role of thermal rearrangement and configurational free volume in plasticization resistance. We demonstrate that glassy polymers exhibiting configurational free volume exhibit better swelling resistance relative to conventional polymers. Finally, comparisons between pure-gas and mixed-gas induced dilation are discussed.

9:30 AM

Oral 11 - Hydrogen- and acetylene-stable silver ion facilitated transport membranes: the effect of salt content on fractional free volume and gas permeability

Matthew N Davenport (The University of Texas at Austin)*; Tangqiumei Song (UT Austin); Justin J Rosenthal (The University of Texas at Austin); Cara Doherty (CSIRO); Joan Brennecke (The University of Texas at Austin); Benny Freeman (The University of Texas at Austin)

Silver ion facilitated transport membranes exhibit high olefin-paraffin selectivity (e.g., ethylene-ethane) but poor stability of the Ag+ carrier, especially in the presence of reducing gases common in dehydrogenation processes, such as H2, H2S, or C2H2. Solid polymer electrolyte membranes comprised of crosslinked poly(ethylene glycol) diacrylate and containing up to 70 wt% silver bis(trifluoromethylsulfonyl)imide (AgTf2N) salt were synthesized as dense films with constant crosslink density. Following 10 weeks of cumulative pure H2 permeation at 4 bar and 35°C, the membrane retained over 90% of its initial ethylene-ethane selectivity—separation performance which surpasses the polymeric gas separation membrane upper bound. The AgTf2N-containing membranes are similarly resistant to C2H2; however, reduced gas permeability and formation of a Ag2S layer were observed following H2S permeation. High salt concentration (i.e., 70 wt%) is needed to obtain the highest selectivity, but increasing cohesive energy density reduces gas permeability and enhances size selectivity. Thus, the effects of AgTf2N concentration on fractional free volume and the size of free volume elements accessible for penetrant diffusion were characterized with Positron Annihilation Lifetime Spectroscopy (PALS). Fractional free volume decreases non-linearly with increasing salt concentration, with a clear transition in physical property trends observed at 50 wt% AgTf2N. This result is discussed in the context of polymer chain dynamics (i.e., Tg), gas penetrant size and condensability, the activated diffusion model, and ion associations and distribution in the polymer matrix. Thin-film composite (TFC) membranes with a 1 μm thick active layer were synthesized as flat sheets coated on a porous polyetherimide support, and the mixed-gas ethylene-ethane selectivity under realistic compositions is reported in preparation for scale-up and pilot-scale testing.

9:50 AM

Oral 12 - Exploring the separation efficiency of ultra-selective 3D CANAL hydrocarbon polymeric membranes for hydrogen-based separations after long-term aging

Jing Ying Yeo (Massachusetts Institute of Technology)*; Francesco Maria Benedetti (Massachusetts Institute of Technology); Zachary Smith (Massachusetts Institute of Technology)

Hydrogen (H2), recognized as a highly efficient energy carrier, is emerging as a pivotal player in global decarbonization. Membrane-based separations are gaining prominence due to their energy efficiency and modularity, making them a promising technology for hydrogen purification. The advent of processable microporous polymers has significantly pushed the performance limits. However, polymer physical aging remains a challenge, leading to a significant decline in gas permeability over time. Our study focuses on 3D CANAL polymers, a class of rigid ladder polymers synthesized through efficient catalytic arene-norbornene annulation (CANAL) polymerization. Previous research demonstrated an unconventional aging behavior of CANAL-Me-S6F, with a remarkable 1000% gain in H2/CH4 permselectivity and only a 33% reduction in H2 permeability after aging for 158 days, surpassing the 2015 Upper bounds. Building on this foundation, our current investigation focuses on the gas transport performance of ultra-aged (>365 days) 3D-CANAL polymers for H2-based separations, exhibiting a total 6100% boost in H2/CH4 selectivity and a 2400% boost in H2/N2 selectivity with a mere 55% loss in H2 permeability. Furthermore, we evaluate the stability of these ultra-aged polymers under industrially relevant binary and ternary gas mixtures and high-temperature conditions. Notably, at longer aging times, H2/CO2 selectivity improves due to increased size-selectivity with simulations predicting up to H2 15000 barrer at 190°C. In conclusion, this study elucidates the robustness and potential applications of ultra-selective hydrocarbon polymers for hydrogen separations. The insights gained pave the way for advancements in clean energy technologies and contribute to the broader landscape of sustainable energy solutions.

10:10 AM

Oral 13 - Evaluation of air addition into methane dry reforming reaction in Pd/YSZ membrane reactor for syngas production

Omid Jazani (Clarkson University)*; Simona Liguori (Clarkson University)

The methane dry reforming (MDR) process is performed to produce synthesis gas (H2 and CO). The reaction is endothermic and operates at 800-1000 °C and 20-30 bar.

CH4 + CO2 → 2CO + 2H2 | ΔH°298 = 247 kJ mol⁻1 (methane dry reforming)

The MDR reaction has issues such as unwanted reactions and constraints of thermodynamics. Unwanted reactions, such as methanation, methane decomposition, reverse water gas shift, and Boudouard reactions lead to carbon deposition which reduces the conversion and hydrogen yield.

The use of a hydrogen-selective membrane reactor (MR), such as Pd-based MR to perform MDR reactions can mitigate the coke deposition issues. Particularly, since the MR selectively removes hydrogen from the reaction zone simultaneously. It can hinder the unwanted methanation and RWGS reactions, which are the precursor of coke and it can lead to an increase in the performance in terms of conversion, hydrogen yield, and recovery.

In this work, the MDR reaction is performed in a Pd/YSZ MR packed with a commercial nickel catalyst to analyze the effect of temperature and pressure variation on hydrogen production and coke formation. In addition, air is introduced in a small percentage to the system to evaluate the effect of oxygen on the MR performance. It is expected that oxygen reacts and consumes carbonaceous materials, limiting the coke issue, and reacts with part of methane by partial oxidation to produce more hydrogen. This should increase the overall MR performance. A comparison of the MR performance in terms of conversion and hydrogen yield with the ones obtained by a conventional reactor is also evaluated and discussed. Finally, the pristine and used membrane is characterized by Scanning Electron Microscope (SEM) and Energy Dispersive X-ray spectroscopy (EDS) technique.

10:30 AM

Oral 14 - Synthesis and Characterization of ZnCl2-Immobilized Molten Salt (IMS) membrane for NH3 separation

Moses T Adejumo (Clarkson University)*; Simona Liguori (Clarkson University)

Ammonia (NH3) production is widely acknowledged for its role in fertilizer manufacturing and emerging use as a hydrogen carrier. Currently, the conventional Haber Bosch (H-B) process for NH3 production generates 1.5 - 1.6 t_(CO2-eq)/t_(NH3-produced). Switching from traditional NH3 separation methods to energy-efficient alternatives like membrane separation can help mitigate CO2 emissions. In this study, a ZnCl2 immobilized molten salt (IMS) membrane is synthesized in situ and evaluated for the potential separation of NH3 from a gas mixture containing N2 and H2. The membrane was prepared via direct deposition technique using woven wire mesh of 1 μm nominal pore size. Several permeation tests were performed to evaluate its permeation characteristics at 300 °C and different pressures, such as 1 - 4.5 bar. For single gas permeation, NH3 permeance was found to be as high as 211 GPU, with both NH3/N2 and NH3/H2 ideal selectivities higher than 107. For binary mixtures, NH3 permeance was as high as ~6700 GPU at ~3 kPa. A model was also developed, which reasonably predicted the facilitated transport of NH3 through the membrane. The mean absolute percentage error between the model and experimental data was 3.9 - 4.2%. Kinetic and thermodynamic parameters associated with NH3 coordination with Zn2+ were determined from the model. The membrane was stable for at least 640 h under different feed mixtures with no significant performance loss. This membrane demonstrates suitability for gas separation in processes featuring low NH3 partial pressures, besides its potential application in the H-B process.

Functional Membrane Fabrication and Environmental Sustainability

8:30 AM

Oral 15 - Scaled Manufacturing of Supported Loose-NF Membranes with Eco-Friendly Solvents for Water Treatment Applications

David Lu (University of Kentucky)*; Kwangjun Jung (Georgia Institute of Technology); Ju Young Shim (Georgia Institute of Technology); Tequila Harris (Georgia Institute of Technology); Isabel Escobar (University of Kentucky)

The lack of sufficient translational research between materials research and module development remains a strong barrier to bringing innovations to the forefront of the membrane field. In the case of polymeric membrane fabrication, the translation of green solvent research is imperative due to the onset of regulations phasing out traditionally used solvents. This study proposes slot die coating (SDC) as a method to scale the manufacturing of polymeric membranes prepared with eco-friendly solvents. SDC was applied to fabricate loose-nanofiltration polymeric membranes were prepared using polysulfone (PSf) and the eco-friendly solvents Rhodiasolv® PolarClean and gamma-Valerolactone (GVL). The membranes were prepared on a porous support layer, which has not been studied in SDC preparation of membranes.

The rheological properties of the PSf-PolarClean-GVL dope solution were analyzed to guide the proper coating parameters. A range of solution flowrate and substrate speed values were used to prepare membrane samples, which resulted in the formation of a coating window that visualizes the parameter combinations that formed membranes with defects and defect-free membranes.

A comparison study between membrane samples prepared using SDC and doctor blade extrusion (DBE) was completed. Pre-wetting the support layer with dope solution minimized shrinkage of the membrane film thickness. SDC samples exhibited properties and filtration characteristics consistent with DBE samples. Pre-wetted samples produced using SDC exhibited the highest rejection of bovine serum albumin (99.20% ± 1.31%) and high mean permeability (70.5 ± 8.33 LMH/bar). SEM images showed exhibited similar pore structures and mean pore sizes between SDC (34.6 ± 10.5 nm) and DBE membranes (43.2 ± 9.1 nm). This study demonstrates that combining sustainable materials development with SDC provides a holistic approach to membrane separations to bridge the gap between new and exciting materials and membrane formation.

8:50 AM

Oral 16 - Sustainable additive manufacturing of polysulfone membranes for liquid separations

Brian Leonard (West Virginia University)*; Oishi Sanyal (West Virginia University); Harrison Loh (West Virginia University); Kostas Sierros (West Virginia University); David Lu (University of Kentucky); Ebuka Ogbuoji (University of Kentucky); Isabel Escobar (University of Kentucky)

Polymer membranes have traditionally been fabricated by phase inversion methods. Technologies such as doctor blading, slot die extrusion and hollow fiber spinning are some of the prior reported technologies used for fabricating large membrane areas that utilize this phase inversion mechanism. The large quantities of solvents needed in these processes are toxic and harmful to the environment. This has led to a desire to utilize novel 3D printing methods, that enable reduction in solvent waste and better control of membrane fabrication parameters. In this talk, I will demonstrate a new direct ink writing (DIW) process for designing polysulfone membranes that significantly minimizes material and solvent usage, while also providing a set of new parameters to create membranes with tailored properties.
I will also provide an analysis of the critical parameters of the DIW process that affect the separation properties of polysulfone membranes – examples of such parameters being substrate type, ambient humidity, and post-processing methods. Transport properties of membranes including permeability and Bovine Serum Albumin (BSA) rejection, in addition to morphological properties and their comparison with doctor bladed polysulfone membranes will be discussed Finally, I will highlight the challenges associated with this new technology and its future prospects as a sustainable membrane manufacturing strategy.

9:10 AM

Oral 17 - An alternative application of 2D carbon-based membranes for efficient air-separation and H2 purification from gas mixtures

Dániel Gardenö (UCT Prague)*; Karel Friess (UCT Prague); Zdenek Sofer (UCT Prague); Malte Kaspereit (FAU Erlangen-Nürnberg)

This proposed work explored the effects of combining two primary 2D carbon-based materials, graphene oxide (GO) and single-walled carbon nanotubes (SWCNT), with various inorganic additives. These novel GO-SWCNT composite membranes were tested for effective H_2 separation from permanent gases such as methane and carbon dioxide. These composites benefit from unlimited permeability-selectivity performance. At the same time, polymer-based commercial organic membranes usually suffer from the trade-off effect, and they also tend not to withstand higher temperatures. On the contrary, inorganic and carbon-based materials usually tend to be offset by their problematic scaling-up, fragility, and complicated preparation. To mitigate unwanted issues, such as fragility and lack of reproducibility, the influence of the SWCNT as support and modification of the GO skin layer by adding selected polyvalent metal ions on the overall separation properties was investigated. The modification was performed by adding selected polyvalent metal ions into the GO solution before filtration to evoke reported interactions of ions with the GO structure. Doped GO sheets connect via intermolecular bonds, making the material more ordered and stable. Prepared composite membranes were further characterized using conventional analytical methods to reveal the effect of the additives on the structure of materials and transport properties. Combining bivalent metal ions with a graphene oxide provided enormous H_2 permeability with excellent CO_2 and CH_4 selectivity, surpassing the corresponding Robeson upper bounds. Interesting results were also obtained for air separation. These results demonstrate the uniqueness of the prepared composite materials as they exceed the vast majority of published membrane materials and further explicit the possibility of molecular engineering level tailoring of the GO layer thickness that leads to tuning the permeability and raises questions about scaling their synthesis up.

9:30 AM

Oral 18 - Understanding the mechanism of additive-assisted isoporous membrane formation through X-ray scattering

Noah P Wamble (University of Texas at Austin)*; Mostafa Nassr (UT Austin); Matthew Landsman (Lawrence Berkeley National Lab); Louise Kuehster (University of Texas at Austin); Joan Brennecke (The University of Texas at Austin); Nathaniel Lynd (The University of Texas at Austin); Gregory Su (Lawrence Berkeley National Lab); Benny Freeman (The University of Texas at Austin)

Self-assembly plus nonsolvent induced phase separation (SNIPS) is a transformative technique where block copolymer self-assembly is combined with NIPS to manufacture ultrafiltration membranes with isoporous pore size distributions. These isoporous membranes could surpass the upper bound for homopolymer NIPS membranes as their narrow pore size distribution produces membranes with both high flux and high selectivity. However, one of the limitations of isoporous membranes is that defects must be minimized to realize the benefit of the narrow pore size distribution. Polystyrene-b-poly(4-vinylpyridine) (SV) is the most prevalent polymer for making isoporous membranes, and poly(4-vinylpyridine) is known to complex with metal salts. Therefore, these metal salt additives are added to casting solutions to form complexes with the polymer to stabilize the solution ordering and preserve the ordered structure during the dry step to minimize the number of defects. In this work, the mechanism through which metal salt additives narrow the top surface pore size distributions of isoporous SV membranes is explored. Differential scanning calorimetry (DSC) shows that the metal additives preferentially bind to the hydrophilic V block through measuring changes in glass transition temperatures. Small angle X-ray scattering (SAXS) and dynamic light scattering (DLS) techniques tracked changes in solution ordering and aggregation as a function of additive loading. Next, in situ grazing incidence SAXS experiments recorded how the order changed during the dry step after casting these SV membranes with and without metal additives. Copper additives are shown to change the ordering in solution by increasing the segregation strength, χN, between the polymer blocks.

9:50 AM

Oral 19 - Elucidating the Structure-Property-Performance Relationship of Incorporating Waste Polymethyl Methacrylate in Polyvinylidene Fluoride Membrane Fabrication

Joseph T Danner (The University of Alabama)*; Liam Papier (The University of Alabama); Steven T Weinman (The University of Alabama)

With an annual production of ~4 million metric tons, polymethyl methacrylate (PMMA) is a widely used polymer due to its optical properties, biocompatibility, and hardness. Estimates indicate that only 10% of waste PMMA is recycled, while the majority is disposed by conventional means. Furthermore, current recycling processes are typically energy intensive and/or produce inferior materials, which presents an opportunity to develop alternative methods to address these downfalls. A promising strategy involves solution blending, where additional polymer(s) are dissolved with PMMA to form a miscible blend. Previously, researchers have identified polyvinylidene fluoride (PVDF) as a compatible polymer to improve the mechanical properties of PMMA for membrane fabrication. This study probes the structure-property-performance relationship of PVDF/PMMA blends to understand how different additive and molecular weight compositions in waste PMMA sources affect PVDF microfiltration (MF) membrane properties. Laboratory cuvettes and fiber-optic cables were used as sources of PMMA waste. 16 wt% PVDF/PMMA blends were prepared with two greener solvents, dimethyl sulfoxide (DMSO) and triethyl phosphate (TEP), to produce flat-sheet MF membranes using nonsolvent-induced phase separation (NIPS). Waste plastics were evaluated with spectroscopic and chromatographic techniques to characterize polymer properties and to determine if processing additives or co-polymers were present. Scanning electron microscopy (SEM) was used to evaluate the cross-sectional and top surface morphologies of fabricated membranes. Pure water permeance and polystyrene (PS) microplastic rejection were quantified with dead-end filtration. We found that membrane hydrophilicity and rejection proportionally increased as the PMMA loading increased from 0 to 8 wt%. In contrast, the permeance and surface porosity exhibited the opposite trend, decreasing from 768 ± 64 to 37 ± 4 LMH/bar and 34 ± 5 to 6 ± 1%, respectively.

10:10 AM

Oral 20 - Enhanced bipolar membrane water disassociation reaction with layer-by-layer assembled clay composite junctions

Nadia Boulif (Eindhoven University of Technology)*; Menno Houben (Eindhoven University of Technology); Zandrie Borneman (Eindhoven University of Technology); Kitty Nijmeijer (Eijndhoven)

Bipolar membranes (BPMs) are gaining increasingly more attention as they are essential for developing and optimizing electrochemical processes necessary to shift towards a more sustainable economy. In this context, it is required to develop BPMs with enhanced mechanical stability, high permselectivity, and low water dissociation resistance (RWD).

In this work, electrospun BPMs with a layer-by-layer (LbL) assembled montmorillonite (MMT) clay-polyelectrolyte (PE) composite junction catalyst are studied. In particular, the focus is on the effect of the bilayer number (three, six, and nine) and the PE strength (poly(ethylenimine) (PEI) as a weak PE and poly(diallyl dimethylammonium chloride) (PDADMAC) as a strong PE) on the BPM performance.

Adding the multilayers in the BPM junction decreases the membrane resistance in reverse bias from 560 Ohms·cm2 to 21 Ohms·cm2 for the best-performing BPM. The bilayer number has limited influence on the overall membrane resistance, while the PDADMAC BPMs outperform the PEI BPMs due to higher PE and clay adsorptions. Electrochemical impedance spectroscopy shows that the junction capacitance increases by 1.4 fold from three to six bilayers for both PEs, since junctions with a larger number of catalytical sites can produce protons and hydroxide ions more quickly. On the other hand, the RWD is more dependent on the PE type. For the same number of bilayers, the RWD of PDADMAC BPMs is 10% higher than the PEI BPMs due to the non-catalytic activity of the PDADMAC quaternary ammonium groups compared to the weak amine groups in PEI, which also function as water dissociation catalysts.

Furthermore, the higher Donnan exclusion at the modified junctions improves the BPM permselectivity by threefold compared to the BPM that contains no catalyst. Thus, adding MMT clay composite LbL catalyst to BPM junctions is a promising method to improve the efficiency and reduce the energy consumption of electrochemical processes that rely on BPMs.

10:30 AM

Oral 21 - Advancing sustainable membrane fabrication by investigating the effect of biodegradable and non-biodegradable solvents on polylactic acid (PLA) membranes

Escobar, Isabel (University of Kentucky)*; Ogbuoji, Ebuka (University of Kentucky)

Global industrialization has led to declining air quality, resulting in an increased need for effective air purification solutions. For critical applications such as healthcare, high efficiency particulate air (HEPA) filters capable of capturing over 99% of particulate matter 0.3 µm or smaller (PM0.3) are recommended by the Centers for Disease Control and Prevention (CDC). While conventional HEPA filters are effective at trapping fine particulate matter, they have some limitations including decreased filtration efficiency over time due to charge depletion and non-biodegradability leading to environmental concerns. Recent studies have focused on developing biodegradable HEPA flat sheet polymeric membranes using polylactic acid (PLA), a biodegradable polymer. However, the solvents used in the fabrication process are usually toxic and highly concerning. This study investigated the effect of substituting commonly used hazardous solvents (NMP, DMAc), with more sustainable alternatives (ETAc, GBL) for fabricating biodegradable flat sheet PLA membranes via non-solvent induced phase separation (NIPS). Fourier transform infrared spectroscopy (FTIR) showed similar surface chemical compositions for all fabricated membranes and pristine PLA. Differential Scanning Calorimetry (DSC) and Thermogravimetric Analysis (TGA) showed traditional toxic solvents demonstrated a stronger plasticization effect on PLA due to their higher potential to form polymer-solvent interactions, as indicated by the calculated relative energy difference (RED) values. Additionally, contact angle measurements showed that the membrane fabricated using ETAc was the most hydrophobic (125 ± 3o), an important feature for air filters which prevents colonization of microorganism due to water accumulation. All membranes showed sufficient air flow and sodium chloride aerosol filtration efficiency (≥ 96%). A 50% biodegradation was observed when the PLA membranes were subjected to proteinase K at pH 8.0 for 3 hours.

Process Intensification and Decarbonization

8:30 AM

Oral 22 - Zeolite Membrane Reactor for CO2 Hydrogenation to Methanol: A Simulation Study on Comparison of Water and Methanol Selective Membranes

Sieun Jung (Korea University/Arizona State University); Mahsa Sharafi (Arizona State University); Jerry Lin (Arizona State University)*

The production of methanol through CO2 hydrogenation enables the conversion of greenhouse gas into a valuable chemical or energy storage medium. However, CO2 hydrogenation reaction exhibits low thermodynamic equilibrium conversion with two side reactions that limit its selectivity for methanol. A water or methanol-selective zeolite membrane reactor allows the separation of either product water or methanol during the reaction, enhancing the conversion of CO2 and, consequently, the yield of methanol. A mathematical model for water and methanol-selective zeolite membrane reactors was established to study their performance in methanol production. The simulation study shows that the membrane properties, membrane reactor, and operational conditions should be selected to ensure that the Damkohler number (the ratio of reaction to flow rate) matches the dimensionless methanol permeance of the membrane. If the water or methanol to H2 or CO2 perm-selectivity is low, the membrane reactor would function as a membrane separator, with reactants permeating through the membrane before being consumed. This results in CO2 conversion significantly lower than the equilibrium value. The operational pressure is a key parameter affecting the zeolite membrane reactor's performance. With a highly permeable and selective zeolite membrane and an active catalyst, both water and methanol-selective zeolite membrane reactors achieve CO2 conversion and methanol yield significantly higher than the thermodynamic equilibrium values. Under optimal conditions, both zeolite membrane reactors can produce methanol at a similarly high methanol yield of up to 80%. However, the methanol-selective zeolite membrane reactor offers higher selectivity than the water-selective zeolite membrane reactor.

8:50 AM

Oral 23 - Catalytic Membrane Reactors for combined separation and contaminant degradation

Nan Zhang (Tianjin University); Yichen Wu (McMaster University); Gavin Yuen (McMaster University); Abdelrahman Awad (McMaster University); Emmanuel Doelsch (Universite Montpellier); Charles-François de Lannoy (McMaster)*

Catalytic membrane Reactors (CMRs) are a new process intensified technology that combine catalysis with membranes by seeding their internal surface area with catalysts. This enables both high flux separation and continuous catalytic degradation of small contaminants, while maintaining nanocatalyst size, stability, and reactivity. Noble metal catalysts have high catalytic activity and fast reaction kinetics, while non-noble catalysts are more practical for large scale applications such as wastewater treatment. We have produced CMMRs containing noble metal Pd-nanocatalysts and those containing non-noble Ni and Ni-Cu bimetallic catalysts. We report a simple 2-step batch reaction approach for synthesizing CMMRs applicable to noble and non-noble metal nanocatalysts alike. Nanoparticle distribution, composition, and structure were verified by TEM and x-ray absorption near-edge spectroscopy (XANES), and supported by XPS. The synthesized CMMRs were evaluated for their ability to transform the environmental contaminant 4-nitrophenol (4-NP) to 4-aminophenol (4-AP). Both noble and non-noble CMMRs exhibited strong catalytic reduction of 4-NP (>99%). Ni-Cu CMMRs boasted a reaction rate constant of 725.03/min, the sources of which were identified as the synergistic electronic effects of the Ni-Cu bimetallic structure, the catalysts' oxidation states, and their high surface area. 4-NP conversion on the Ni-Cu catalysts followed the Langmuir-Hinshelwood mechanism, and conversion efficiency was dependent on flow rate, representing a critical optimization trade-off. By combining membrane filtration and catalysis, CMMRs showed both >99% 4-NP conversion and >90% 1 MDa PEO rejection, demonstrating the ability for these membranes to separate conventional contaminants at fluxes similar to MF membranes, while continuously targeting and degrading small organic molecules, demonstrating exciting possibilities for use in industrial wastewater treatment.

9:10 AM

Oral 24 - Computational Fluid Dynamics and Systems Engineering to minimize membrane-based carbon capture costs

Hector A Pedrozo (Carnegie Mellon University)*; Cheick Dosso (Carnegie Mellon University); Thien Tran (National Energy Technology Laboratory); Lingxiang Zhu (National Energy Technology Laboratory); Victor A Kusuma (NETL); David Hopkinson (NETL); Lorenz Biegler (Carnegie Mellon University); Grigorios Panagakos (Carnegie Mellon University)

Carbon capture technologies offer promising solutions for mitigating CO2 emissions in existing facilities. Among the available technologies, membrane-based carbon capture stands out due to its notable attributes; energy-efficient, flexibility, and rapid response times, all contributing to reduce operating costs and a smaller carbon footprint. Despite its potential, widespread adoption encounters challenges related to both capital costs and membrane stability.

To overcome these difficulties and facilitate scaling-up of membrane technology, high-fidelity computational fluid dynamics (CFD) simulations have become a valuable tool. These simulations offer a detailed understanding of the fluid mechanics and mass transfer within membrane systems. In addition, achieving high CO2 recovery and product purity ratios, necessitates the implementation of a multi-stage membrane process, prompting a thorough exploration of optimal membrane network configurations. To tackle design challenges, membrane systems engineering has been developed.

In this study, we propose a superstructure for membrane-based carbon capture, addressing these challenges through the formulation of an optimization problem. The case study considers a flue gas with a flowrate of 1000 mol/s and a 30 % CO2 (molar basis), which is representative of steel plants. Reduced models, derived from rigorous CFD simulations of membrane modules, are utilized to enhance the efficiency of the optimization process. The reduced model includes specific input variables such as inlet flow rate, retentate and permeate pressures, and inlet CO2 concentration; while the CO2 recovery and purity in the permeate side are considered as output variables. In this way, we exploit the information from the CFD simulations in the superstructure, aiding in the selection of the optimal configuration for the multi-stage membrane process. The results underscore the efficiency of a multi-stage membrane design, showing capture costs around 45 $/t-CO2.

9:30 AM

Oral 25 - Innovating Membranes for Carbon Capture: From Material Synthesis to Process Prototype

Yang Han (The Ohio State University)*; Yutong Yang (The Ohio State University); Jingying Hu (National Renewable Energy Laboratory); Ruizhi Pang (The Ohio State University); Babul Prasad (The Ohio State University); Winston Ho (OSU)

Sustainable energy and environment, to a large extent, hinge on the acquisition, capture, and utilization of small molecules such as CO2. As an energy-efficient separation technology, polymeric membranes have gained traction for carbon capture. Yet, designing them to efficiently handle small molecules with minimal size differences remains a notable challenge, especially in terms of achieving adequate gas selectivity using scalable membrane materials. This presentation will focus on our work in designing facilitated transport membranes (FTMs), where carriers are incorporated to mediate the diffusion of reactive penetrants over inert ones. In the first part, the molecular engineering of FTMs employing amines and amidines as carriers for CO2 capture will be discussed. This resulted in the development of a high-performance membrane with a CO2 permeance of 4,200 GPU and a CO2/N2 selectivity of 165. The second part will describe the technology upscaling process, progressing from membrane coupons to roll-to-roll continuous membrane fabrication, and ultimately to membrane module fabrication. Subsequently, the construction of a process prototype based on a two-stage enriching cascade design will be outlined. Field test data with actual flue gases will also be presented to address issues regarding membrane scalability, operability, and stability. Finally, the calibration and validation of the process model using field test data will be discussed, which facilitated the adjustment of operating conditions to accommodate varying flue gas compositions and CO2 recovery needs. From molecular to process level, our endeavors aim to pave the way towards deployable membrane solutions for carbon capture.

9:50 AM

Oral 26 - Advancements of Responsive and Catalytic Membrane Reactor for Water and Halo-Organics Remediation

Dibakar Bhattacharyya (U. Kentucky)*; Rollie Mills (U. Kentucky); Francisco Leniz (U. Kentucky); Sam S Thompson (University of Kentucky); Subrajit Bosu (U. Kentucky)

The development of multifunctional and responsive membranes for environmental related process intensification requires a high level of control of both the characteristics of the base polymeric or inorganic support layer, and the corresponding surface properties. The incorporation of nanostructured catalytic materials in microfiltration (MF) type functionalized membranes or creating metal-polymer composite membranes reduces unit operation steps can in the area of pollutant remediation from water. In addition, the use of macromolecules functionalization in MF type membrane pores, such as, poly-acrylic acid (PAA), poly-glutamic acid (PLGA) provide pH responsive behavior through helix-coil transitions, whereas poly-N-isopropylacrylamide (pNIPAAm) provides temperature responsive behavior from hydrophilic to hydrophobic domain. We will demonstrate that this property can be used to adsorb/desorb PFAS compounds. Catalytic metals can be easily be synthesized in membrane pores through ion exchange and post reduction steps (Fe/Pd for example) or one can deposit thin catalytic metal films on MF/UF surface through magnetron sputtering technique. We demonstrated the use of magnetron sputtering to deposit thin catalytic films on top of porous membranes. A dealloying step in water removes the Mg component of the film creating a nano-porous (np) film of Pd for organic dechlorination reactions from water in the presence of dilute hydrogen gas. The presentation will include: (1) pore functionalization approaches to create temperature and pH responsive membranes with demonstration of pollutant sorption/desorption (such as, PFOA), (2) Further enhancement of PFAS separations by integrating an NF layer on temperature responsive MF membrane, (3) catalytic dehalogentation of halo-organics through pore functionalized Fe/Pd nano-catalyst, (4) process intensification aspects involving membrane functionalization and membrane reactor. This research has been supported by NIEHS, and NSF.

10:10 AM

Oral 27 - Intensified propane dehydrogenation in carbon molecular sieve hollow fiber membrane reactors

Lu Liu (University of Maryland); Antara Bhowmick (University of Delaware); Dongxia Liu (University of Delaware); Chen Zhang (University of Maryland)*

Propylene is an important bulk chemical with global annual production exceeding 100 million tons. Non-oxidative propane dehydrogenation (PDH) is increasingly practiced to address the gap between propylene demand and production. High reaction temperatures (above 600 °C) are needed to overcome thermodynamic limitations and to provide high propane conversion, which not only gives large CO2 emissions but also leads to rapid catalyst deactivation. Membrane reactors can enhance propane conversion above equilibrium conversions by selective hydrogen removal according to the Le Chatelier's principle. This can give commercially attractive propane conversion at lower reaction temperatures with reduced CO2 emissions.

In this talk, we will present novel carbon molecular sieve (CMS) hollow fiber membrane reactors for intensified and decarbonized propane dehydrogenation and product separation.(1) The membrane reactor consists of asymmetric CMS hollow fiber membranes (diameter of 300 micron) for hydrogen removal/separation and platinum-zeolite catalysts for low-temperature propane activation. The CMS hollow fiber membranes were made by pyrolysis of polyimide precursor hollow fibers and had a combination of attractive hydrogen permeance and hydrogen/hydrocarbon separation factor at 400-600 °C. With fast and selective hydrogen removal/separation, the membrane reactor showed propane conversions that were 300% above the equilibrium conversion. This gave outstanding propane conversion (up to 45%) and high propylene selectivity (above 95%) at 450 °C, which is 150 °C lower than most commercial PDH reactors.

10:30 AM

Oral 28 - Mathematical Modeling & Optimization of a Water-Gas Shift Membrane Reactor for Low-Cost H2 Production with CO2 Capture

Hani A.E. Hawa (Precision Combustion, Inc.)*; Kyle Hawley (Precision Combustion, Inc.); Damian T Agi (University of Notre Dame); Alexander Dowling (University of Notre Dame)

Water-gas shift membrane reactors (WGS-MRs) are promising pathway for the production and extraction of low-cost H2 from renewable feedstocks, including biomass. Metal-based membranes, including Pd and its alloys, have been extensively studied due to their infinite permselectivity towards H2 and suitability for high-temperature operation. To exploit their cost benefits for H2 production with simultaneous CO2 concentration, WGS-MRs performance needs to be optimized. This includes tuning the complex set of operating conditions (temperature, pressure, space velocity, sweep) and performance parameters (permeance, recovery), and material properties (permeability) at relevant process scales.

We developed a mathematical modeling framework for a Pd-based WGS-MR that accounted for reaction kinetics and mass transport. The model was implemented in Pyomo, an open-source algebraic modeling language, and was qualitatively validated by comparison with lab-scale data at representative conditions.

We demonstrated the optimization capabilities of our model by examining multiple variables to maximize H2 recovery and minimize production cost. Further, we evaluated the economic viability of the WGS-MR for the production of blue H2 from gasification of mixed feedstocks through techno-economic analysis (TEA). Our TEA was benchmarked against the H2 Production Model (H2A) developed at the NREL. Our preliminary TEA showed that replacing the two-stage WGS reactor and pressure swing adsorption units in the H2A process with WGS-MR boosted H2 recovery and resulted in cost savings and heat integration opportunities.

The WGS-MR model developed in this work serves as a design and performance optimization tool to inform the integration of membrane reactors in blue H2 production. It could be adapted to other membrane reactors for process development. It also supports integration into open-source process flowsheet modeling environments, such as DOE NETL's IDAES-PSE framework.

Abstracts – Oral Presentations, Monday Afternoon

Presenting authors are indicated by an asterisk*.

Water Treatment and Desalination Processes

1:45 PM

Oral 29 - Revisiting the value of high-flux seawater reverse osmosis membranes with variable-flow energy recovery

Sangsuk Lee (UCLA)*; Eric Hoek (UCLA)

Despite significant advancements in membrane technology and process design, seawater reverse osmosis (SWRO) remains energy-intensive with state-of-art specific energy consumption (SEC) at or above 2.2 kWh/m3. We introduce a new type of energy recovery device (ERD), the variable flow-pressure exchanger (VF-PX), which combines the high efficiency of traditional isobaric ERDs with the capability to handle variable flow rates. Our thermodynamic analysis compares the VF-PX with conventional ERD systems across various product water recoveries. We demonstrate that the VF-PX can achieve a SEC as low as 1.8 kWh/m^3 at a water recovery rate of 60%, outperforming existing technologies in the high recovery range. Additionally, we explore the impact of enhanced SWRO membrane water permeability on SEC in two-stage systems using different ERD configurations. Our results reveal that while conventional PX systems require very high water permeability near to 9 LMH/bar to achieve 1.8 kWh/m3 at a water recovery of 60%, VF-PX systems can bring the SEC with a water permeability of 5 LMH/bar. The VF-PX shows remarkable energy-saving potential by fully utilizing the residual pressure in the brine stream from the second stage, a feature unique to its variable flow capability. This study underscores that the VF-PX system can significantly benefit from increasing water permeability of membrane in two-stage SWRO with high water recovery. The synergy between high-flux membranes and the novel VF-PX ERD will offer a path towards more sustainable, efficient, and widely accessible SWRO desalination solutions. This technology could potentially lower the SEC for high water recovery applications, marking a significant step forward in membrane-based desalination processes.

2:05 PM

Oral 30 - Performance and Environmental Evaluation of Biochar-based Membranes Synthesized from Various Petroleum- and Bio-derived Solvents

Abelline K Fionah (University of Kentucky)*; Isaac Oluk (University of Kentucky); Isabel Escobar (University of Kentucky)

Water contamination resulting from coal spills is one of the largest environmental problems affecting communities in the Appalachia Region of the United States. This coal slurry contains potentially toxic substances, whose leakage into water bodies can lead to adverse health effects not only on freshwater bodies and plant life but also on humans. In this project, two studies were conducted with the goal of removing organic contaminants from the waters of the Appalachian Region collected from Martin County, Kentucky. One study involved the use of biochar to create a biochar-polysulfone (BC/PSf) flat sheet membrane to remove organic contaminants, and the other compared bioderived (Gamma-valerolactone – GVL, and Rhodiasolv® PolarClean – PC) and petroleum-derived solvents (i.e., N-methyl-pyrrolidone – NMP) to improve biochar-polysulfone membranes. The resulting membranes were tested for their efficiency in removing both positively and negatively charged organic contaminants from water. A comparative life cycle assessment (LCA) with accompanying uncertainty and sensitivity analyses was carried out to understand the global environmental impacts of incorporating biochar, NMP, GVL, and PC in the synthesis of PSf/NMP, BC/PSf/NMP, PSf/GVL, BC/PSf/GVL, PSf/PC, and BC/PSf/PC membranes of 1000 m2 surface area. The results showed that the addition of biochar to the membrane matrix increased the surface area of the membranes and improved both their adsorptive and mechanical properties. The environmental impacts normalized to the BC/PSf/GVL membrane showed that the addition of biochar increased global warming impacts, eutrophication, and respiratory impacts by a substantial amount in the membranes, while it had minimal influence on carcinogenic impacts. The environmental impacts were highly sensitive to biochar addition (Spearman's coefficient >0.8). The

BC/PSf membrane with Rhodiasolv® PolarClean had the lowest environmental impact among all the membranes with biochar.

2:25 PM

Oral 31 - High Recovery Desalination of High Mineral Scaling Source Water via Integrated RO-NF Concentrator, Concentrate Desupersaturation and Second Stage RO-NF Desalting

Yakubu Jarma (University of California, Los Angeles); Jose Morales (University of California, Los Angeles); Phoebe Strauss (University of California, Los Angeles); Yoram Cohen (Professor)*

High recovery desalting of high salinity inland water of high mineral scaling propensity is typically limited by calcium carbonate and/or calcium sulfate scaling. Feed dosing with antiscalants enables recovery up to ~65-75% for feedwater of ~12,00-16,000 mg/L total dissolved solids (TDS). However, increasing recovery to levels above about 90% require concentrate desupersaturation (with respect to the target mineral scalants), followed by secondary membrane desalting. Accordingly, the above approach was evaluated for desalting high salinity California San Joaquin and Salinas Valleys brackish groundwater. First stage desalting was carried out via RO desalting in the mode of concentrate recycle. Subsequently, accelerated seeded precipitation (ASP) was employed for desupersaturation of the first stage RO concentrate and for salt harvesting. The above process step, which is critical unit in the overall high recovery desalting process, was evaluated in a mixed tank crystallizer. The induction time for crystallization (for both calcite and gypsum) and crystallization kinetics were evaluated using inorganic seeds of different sizes (and thus different levels of seeds surface area per unit volume) and surface chemistry. ASP kinetics was followed via monitoring of calcium concentration and formed crystals and growth patterns. While the crystallization induction time was correlated with the seeds surface area per unit volume, surface chemistry also had a major role, demonstrating that gypsum seeds were significantly more effective in accelerating both calcite and gypsum crystallization relative to the use of calcite and silica seeds. Secondary desalting of the desupersaturated concentrate (with respect to calcite and gypsum) was assessed to enable overall recovery that surpassed 93%. Finally, we note that technoeconomic analysis indicated that ultimate concentrate disposal cost constitute a significant portion of the overall operational expenses.

2:45 PM

Oral 32 - Electrifying membrane distillations: Localized Joule heating of laser-induced graphene membranes

Md Hasib Al Mahbub (Texas Tech University)*; Mahdi Malmali (Texas Tech University)

Membrane distillation (MD) is an emerging desalination technique that can treat hypersaline brine to produce pure distillate. However, MD's widespread application is yet hindered by low energy efficiency and scaling. Here, I will report on a novel, single-step fabrication method to develop electroconductive (EC) graphene membranes. The surface of these membranes can be heated using Joule heating. Thus, we can apply the heat at the interface of the mass transfer. This approach promises potential improved energy efficiency and reduced scaling by mitigating concentration and temperature polarization.

The EC membranes developed in our work are fabricated by laser irradiation of a polyethersulfone (PES) in raster mode lasing, which converts the surface of the PES into graphene sheets, also known as laser-induced graphene (LIG). Upon formation of the LIG on the membrane surface, it was coated with a thin SiO_2 @ polydimethylsiloxane (PDMS) layer to obtain superhydrophobic surface for MD application. Results indicate the formation of a superhydrophobic (152°) surface on the PES-LIG membrane. For surface heating with the Joule heating technique, direct (DC) and alternating current (AC) were used with different input power ranging from 0.5 to 10W. The thermal response of PES-LIG membranes was recorded with an FLIR camera to find the optimum heating parameters to evaluate MD performance. Surface temperatures in the range of 50-350 °C were recorded for 0.5-10 W input power. The highest average distillate flux achieved was 8.74 LMH (10W and 60 Hz) with AC and 6.34 LMH (10W) with DC Joule heating. For both cases, more than 99% rejection of NaCl was observed, indicating the successful performance of the EC PES-LIG membranes to sustain the mass transfer interface on the MD feed side. Our observation concluded that AC would be preferable over DC as polarity switching in AC significantly reduces electrode corrosion, which is essential for the long-term operation of MD.

3:05 PM
Oral 33 - The analytic solution of interfacial concentration with observed rejection ratio during dead-end membrane filtration
Albert Kim (U. of Hawai'i)*
In this analytical study, we revisit the fundamentals of constant-flux dead-end filtration, develop an exact analytical solution of the interfacial concentration as a function of dimensionless time and observed rejection ratio, and compare the solution with previous work developed for constant intrinsic rejection ratio. The excessive concentration, analytically obtained, consists of three nonlinear terms with time and reaches an asymptotic limit. The Robin (mixed) boundary condition is applied on the membrane surface, and the Dirichlet boundary condition is used at a far feed–entrance. The historical mathematical difficulties for the inverse Laplace transform are resolved using a linear combination of the Laplace transform of error and complementary error functions and applying the convolution theorem. Besides, we analytically obtain the unsteady variation of the interfacial concentration after the pressure release using the global mass balance. The required time to reduce the interfacial concentration to a specific limit was numerically calculated. More importantly, a novel relationship between observed and intrinsic rejection ratios is found and verified using experimental data from the literature.

3:25 PM
Oral 34 - High-Throughput Experimentation for High-Pressure Reverse-Osmosis Membranes
Hongchen Wang (University of Toronto)*
High-Pressure Reverse-Osmosis (HPRO) is a desalination method that utilizes high pressure to force water through a porous membrane, which is recognized to be more energy-efficient than conventional thermal desalination methods. However, HPRO membranes can be prone to rupturing when pressure exceeds maximum thresholds, leading to a loss of solute rejection. To enhance their mechanical performance, it is essential to investigate appropriate material choices and process parameters for the support layer. Non-Solvent-Induced Phase Separation (NIPS) is the preferred method for synthesizing the support layer, but it involves various parameters. Exhaustive manual experimentation can be time-consuming and prone to human error, highlighting the need for an automated optimization strategy to expedite the optimization process in enhancing membrane performance.
This presentation will introduce an ongoing project that aims to develop a High-Throughput Experimentation (HTE) approach to optimize the synthesis process of the support layer for HPRO membranes. The machine layout comprises a liquid handling robot that automates the synthesis process, and a robot arm that transfers the synthesized membranes to the compression tester for mechanical testing. The compression tester assesses the mechanical properties of the synthesized membranes, and an active learning algorithm analyzes the results and determines the subsequent set of experiments. By expediting the optimization process with fewer trials, the HTE approach enables the identification of optimal parameter combinations that offer high mechanical performance with enhanced reproducibility. To date, the system has successfully produced membrane support layers through complete automation. The presentation will focus on the optimization process and share initial results from this automated membrane fabrication.

Advanced Materials for Gas and Vapor Separation II

1:45 PM
Oral 35 - What Membranes Do We Need for Industrial Propylene/Propane Separations?
Yang Liu (MTR)*; Tim Merkel (Membrane Technology & Research Inc.); Richard Baker (Membrane Technology & Research Inc.)
Membranes are attractive alternatives for enhancing or replacing highly energy-intensive cryogenic technologies in olefin/paraffin separations. Despite the extensive R&D efforts in developing advanced

membrane materials and technologies in the past decades, no membrane has been commercialized for this important application. Many types of membranes have been tried, including polymeric membranes, facilitated transport membranes, inorganic membranes, and mixed matrix membranes, but each has advantages and drawbacks. Indeed, developing such a membrane is one of the most challenging topics in the membrane community, requiring close collaboration between academia and industry. In this presentation, we will talk about the necessary membrane characteristics and corresponding processes for industrial propylene/propane separations based on process simulations and our lab and field tests. We will also present results for new polymeric thin-film-composite (TFC) membranes and modules tested under industrially relevant conditions and discuss opportunities for future improvements. Meanwhile, a new empirical mixed gas tradeoff plot for polymeric membranes under industrially relevant conditions is proposed for guiding future membrane developments. We hope this talk can promote advanced material developments that will lead to commercial membranes for propylene/propane separations.

2:05 PM

Oral 36 - Scale-up of perfluoro(butenyl vinyl ether) (PBVE) and perfluoro(2,2-dimethyl-1,3-dioxole) (PDD) copolymers for the separation of R-410A

Abby N Harders (University of Kansas)*; Luke Wallisch (University of Kansas); Michael Lundin (University of Kansas); Chloe Le (California State Polytechnic); Gabby Zaher (University of Washington); Whitney White (Chromis Technologies); Mark Shiflett (University of Kansas)

The separation of hydrofluorocarbon (HFC) refrigerant mixtures is a pressing issue due to environmental legislation regulating their use and production and requiring reclaimed refrigerant to be used in new applications. There is currently no technology deployed in the refrigerant industry for separating azeotropic refrigerant mixtures and energy efficient solutions are needed to avoid wasted refrigerant value and mitigate environmental consequences. Perfluoropolymers have been explored in the literature for HFC separation with small-scale membrane films, but industrial scale applications require larger surface area to volume for efficient and cost effective separations. This study considered the development of hollow fiber membrane modules produced using polypropylene supports coated with a selective layer of perfluoro(butenyl vinyl ether) (PBVE) and perfluoro(2,2-dimethyl-1,3-dioxole) (PDD). A coating apparatus was developed to provide sub-micron coatings of a selective layer of 30 mol% PBVE-co-70 mol % PDD on hollow fiber supports. Correlations for coating thickness as a function of polymer concentration and velocity were developed using Landua's law. SEM imaging was used to verify coating thickness and uniformity. The coated fibers were assembled into membrane modules and permeability and selectivity were measured for the separation of R-410A, a 50-50 wt% mixture of difluoromethane (HFC-32, CH_2F_2) and pentafluoroethane (HFC-125, CHF_2CF_3). Preliminary results indicate that a high purity of HFC-32 (> 95 mol%) can be achieved in a single pass. Future work is focused on optimizing operating parameters to maximize purity, stage cut, and HFC-32 recovery.

2:25 PM

Oral 37 - Selective photonic gasification of strained oxygen clusters on graphene for overcoming the trade-off between pore size and density

Luc S Bondaz (Ecole polytechnique fédérale de Lausanne)*; Anshaj Ronghe (Indian Institute of Science); Shaoxian Li (Ecole polytechnique fédérale de Lausanne); Kristians Cernevics (Ecole polytechnique fédérale de Lausanne); Jian Hao (Ecole polytechnique fédérale de Lausanne); Oleg Yazyev (Ecole polytechnique fédérale de Lausanne); Ganapathy Ayappa (Indian Institute of Science); Kumar Varoon Agrawal (Ecole polytechnique fédérale de Lausanne)

Incorporating Å-scale single-digit pores with controlled size into single-layer graphene has been inherently challenging due to the restricted understanding of pore formation at the atomic-scale. Pore generation via oxidation route has led to Å-scale pore incorporation into graphene lattice. Nevertheless, for enhancing pore incorporation resolution, a fine control over pore evolution from the pore precursor (also called oxygen cluster) is highly attractive. We introduce a novel photonic gasification method of oxygen clusters to form gas sieving pores. First, we demonstrate that oxygen clusters grow into a core/shell structure formed by an ether core surrounded epoxy group in a bid to minimize the core lattice strain. Subsequently, we selectively gasified the core of the cluster at room temperature, using 3.2 eV light, leading to pore formation. Using O3 oxidation temperature as a leverage, we demonstrate the feasibility of maintaining a narrow pore size distribution, independently of pore density. Such results enabled us to simultaneously increase gas permeance and gas pair selectivity, against a common tradeoff and long-lasting challenge in the field. Ultrahigh H2 permeance of 12000 GPU alongside very attractive H2/N2 selectivity of 50 was achieved. We also demonstrated that such pores could selectively sieve CH4 from N2, acknowledged to be an arduous separation. Dedicated molecular dynamics revealed that CH4 permeates with a significantly lower free energy barrier comparing to N2. By being fast (only 5s) and highly uniform, we also demonstrate that photonic gasification can also be scaled up for CO2 capture with CO2 permeance of 1800 GPU alongside attractive CO2/N2 selectivity about 18.

2:45 PM

Oral 38 - Pall Filtration is Enabling the Energy Transition

Soondeuk Jeung (Pall Corporation)*, Amanda Kite (Pall Corporation); Kyle Kathan (Pall Corporation); Varun Vakharia (Pall Corporation); Maria Lingerfelt (Pall Corporation)

The European Green Deal is a plan to achieve carbon neutrality and sustainability in the EU by 2050 by reducing greenhouse gas emissions. After the Green Deal was presented by the European Commission, other countries have joined in by setting emission goals. One of the main challenges of the Green Deal is to decarbonize the energy sector and shift from fossil fuels to renewable sources. Pall filtration technology will play an important role in enabling the energy transition for green hydrogen production and decarbonization processes – both of which are key components in the path to carbon neutrality. Green hydrogen is hydrogen produced from water electrolysis using renewable electricity, such as solar or wind power. Decarbonization requires efficient and robust filtration and separation solutions to remove contaminants and impurities from the gas streams and protect the downstream equipment. This presentation will provide an overview of how Pall is providing liquid/gas separations and solids removal for green hydrogen production and CCUS (carbon capture and utilization or storage) processes.

3:05 PM

Oral 39 - New class of imidazolium-based ionenes for gas separation membranes

Sudhir Ravula (University of Alabama)*; Jason Bara (University of Alabama)

Limiting greenhouse gas emissions, particularly CO2 from existing fossil fuel-fired power plants, stands as one of the greatest challenges of our time. In the United States, coal-fired power plants account for over 50% of electricity generation, with approximately 40% of CO2 emissions originating from post-combustion processes. While amine-absorption technology is widely used for capturing CO2 from post-combustion flue gas (CO2/N2), which significantly escalates the cost of power production. Recognizing the urgent need for more cost-effective solutions, membrane technology has garnered considerable attention for its potential in gas separations, owing to its energy-efficient separation process. In the literature, an extensive array of polymer materials has been explored as selective layers for gas separation. But to achieve high CO2/N2 selectivity, membranes must exhibit favorable interactions with CO2 to enhance CO2/gas solubility selectivity.

Our research group has been pioneering the development of ionenes, a subset of charged polymers where ionic moieties are embedded within the polymeric backbone rather than attached as pendant groups. These ionenes boast excellent mechanical properties for membrane formation and the capacity to hold additional ""free"" ionic liquids (ILs) within their structures. While porous induced microporosity and poly(ether ether ketone) offer distinct advantages, there has been limited exploration into integrating into the ionic backbone. In this study, we present our innovative design and progress in developing a new class of ionene hybrids, the effects of spacing and sequencing of functional features, and related printed materials for gas separation membranes.

3:25 PM

Oral 40 - Adjustable microstructure design of a polyphthalonitrile precursor and the derived novel carbon molecular sieve membrane for gas separation

Chun-Po Hu (Nanyang Technological University)*

Carbon molecular sieve (CMS) membrane is regarded as the next-generation membrane for energy-efficient gas separation. Developing microstructurally adjustable polymeric precursors holds promise for obtaining high-performance CMS membranes. Polyphthalonitrile (PPN) is an emerging candidate for microporous carbon owing to its flexibility in synthesis chemistry, high char yield (~70%), and adjustable microstructure design. A cross-linkable PPN precursor with thermally adjustable polymer chain arrangement was spin-coated on a porous substrate followed by curing up to 400 °C and pyrolysis at 750 °C to obtain the supported PPN-CMS membranes. Notably, the polymer chain arrangement of PPN can be adjusted via curing temperature, leading to an enlarged intermolecular distance along with an ordered aromatic stacking region. As a result, the gas permeability of the PPN membrane cured at 400 °C improved by 10 times compared to that of the counterpart cured at 280 °C; meanwhile, the CO2/N2 and O2/N2 selectivities of the membrane also improved by 24% and 43%, respectively. More importantly, the polymer chain arrangement of the PPN precursor largely affects the microstructure characteristics of the PPN-CMS membrane. This study proposes a hypothetical mechanism for the microstructure evolution from PPN to the derived PPN-CMS membranes. The PPN-CMS membrane obtained from the precursor with a larger intermolecular distance and a more ordered polymer chain arrangement exhibits a smaller ultramicropore size and narrower distribution. Consequently, the PPN-CMS membrane showed promising CO2 and O2 permeabilities of 3568 Barrer and 760 Barrer, respectively, and CO2/N2 and O2/N2 selectivities of 58.7 and 12.5, respectively, surpassing the Robeson upper bounds. This study elucidates the microstructural relationship between the PPN precursors and the PPN-CMS membranes. We believe that PPN with facile microstructure adjustability can be a new class of CMS membrane materials for gas separation.

Hollow Fiber Membranes

1:45 PM

Oral 41 - Experimental and theoretical evaluation of feed flow collar design for shell fed hollow fiber membrane modules

James Tran (UToledo)*, Thien Tran (National Energy Technology Laboratory); Maya Schuchert (Columbia University); Victor A Kusuma (NETL); Lili Sun (University of Toledo); Comfort Oluleke (University of Toledo); Nathan Diemler (NETL); Lingxiang Zhu (National Energy Technology Laboratory); Grigorios Panagakos (Carnegie Mellon University); Glenn Lipscomb (University of Toledo, Toledo)*; David Hopkinson (NETL)

Membrane separation processes are a promising technology for carbon capture from concentrated point sources. Optimal process performance requires membrane module designs which provide counter-current contacting between the feed and permeate streams and allow for use of a permeate sweep. Large hollow fiber modules require the introduction and removal of one stream in the shell space outside the fiber bundle. To achieve near counter-current contacting, module features such as flow distribution collars have been proposed in the literature. However, studies of these features are lacking in the literature.

An experimental and theoretical study of module collar design is presented here. Hollow fiber membranes are prepared by dip coating a poly(vinylidene) (PVDF) support with a polydimethylsiloxane (PDMS) gutter layer and a Pebax 2533 selective layer. Fiber bundles with a well-defined fiber packing are prepared using a 3D printed module. A parallel fiber bundle consisting of 4-9 uniformly spaced fibers is created with printed tabs that align the fibers and create a tubesheet. The tabs are sealed within a printed case that possesses a series of external ports for gas introduction and removal. Uniquely, both port location and the use of a collar to assist fluid distribution in the shell can be varied for the same fiber bundle. Experimental measurements are compared to computational fluid dynamics (CFD) simulations. The experimental module design allows high-fidelity representation of the fiber bundle and module case in the simulations.

Comparisons between experiment and simulation are in good agreement over a broad range of experimental conditions. The detrimental effect of having ports located too close, leading to stagnation regions, is captured as well as the beneficial effects of using a collar for shell-side fluid distribution around the fiber bundle. Such results help validate the use of CFD to develop high-performance module designs.

2:05 PM

Oral 42 - Development of a modular device for evaluating hollow fiber membrane permeance

Timothy Warner (McMaster University)*; Charles De Lannoy (McMaster University); Yichen Wu (McMaster University)

Hollow fiber polymer membranes are becoming the dominant membrane technology in North America and Europe across several sectors including tertiary water treatment, bioreactors, and industrial oily-water separation. Evaluating the permeance and flux of hollow fiber membranes typically involves testing at the module scale, necessitating potting a considerable number of fibers and fitting them into a module. This process is both time-consuming and costly and is usually only done once, i.e. at the beginning of a membrane's life. Appraising the performance of hollow fiber membranes is thus challenging and most labs and operators seeking information about the impact of operating conditions on membranes test flat sheet membranes instead. As such there is a lack of data regarding the performance of hollow fibers under real operating conditions. We have developed a hollow fiber testing device which can test the permeance of multiple individual fibers simultaneously, in an outside-in and inside-out configuration. Nine individual fibers can be installed into the chamber at a time and the membrane permeance can be assessed for individual membranes, or on the permeate flow of sets of 3 or 9 fibers. The selected methodology offers flexibility, simplified testing procedures, and reliable statistical insights into membrane permeance. Permeate flux is evaluated through mass measurement using parallel data logging scales, providing precision at the sub-second level across the fibers. Preliminary validation has demonstrated consistent outcomes across all nine fiber testing ports. This apparatus proves ideal for scrutinizing alterations in the permeance of individual hollow fibers pre- and post-fouling, as well as investigating the impact of various cleaning regimes on the restoration of membrane permeance.

2:25 PM

Oral 43 - Physical Aging of Defect-Free Asymmetric Functionalized PIM Hollow Fiber Membranes

Philippe Jean-Baptiste (Massachusetts Institute of Technology)*; Zachary Smith (MIT)

Polymer hollow fibers represent an important form factor for membranes towards energy-efficient gas separations and organic solvent nanofiltration. These membranes are favored over traditional flat sheet membranes due to their increased surface-area-to-volume ratios, thin selective layers, and greater mechanical stability. Despite their versatility, polymer membranes have limited industrial deployment because of their inherent permeability–selectivity trade-off defined by the Robeson upper bound. To overcome this challenge, polymers with inefficient packing structures and high fractional free volume known as polymers of intrinsic microporosity (PIMs) have been developed. These polymers, specifically PIM-1, can be post-synthetically modified via solid-state functionalization by targeting its nitrile groups, thereby increasing their gas separation performance for certain gas pairs. However, as a result of their excess free volume, membranes made from glassy polymers suffer from a phenomenon known as physical aging. Physical aging represents the relaxation of polymer chains towards an equilibrium state that decreases the material's fractional free volume and permeability over time. In this work, high molecular weight PIM-1 was spun into hollow fibers that underwent solid-state functionalization to yield an amine-functionalized derivative of PIM-1 known as PIM-NH2 to determine the effect of hydrogen bonding and chain mobility on the physical aging of hollow fibers. It was found that, as a result of their thin selective layers, both types of PIM fibers see accelerated aging when compared to dense films. PIM-1 fibers specifically see greater gains in selectivity as permeability decreases over time, likely due to chain alignment during spinning. This work represents the first proper aging study performed on hollow fibers spun from microporous organic polymers, allowing for a better understanding of physical aging and long-term performance of PIM-based hollow fibers.

2:45 PM

Oral 44 - Improved nanofiltration through inverted membrane structures

Tjerk R Watt (University of Twente)*; Wiebe M. de Vos (University of Twente); Joris de Grooth (Univeristy of Twente); Esra te Brinke (University of Twente)

Recently, the accumulation of organic micropollutants (OMP) in ground and surface waters, caused by the disposal of insufficiently treated wastewater streams, has become an increasing concern for human and environmental health and safety. One of the techniques that is being looked at to clean such streams are polyelectrolyte multilayer membranes (PEMM). {te Brinke et al, 2020}

Currently, commercial PEMMs are made in the hollow fibre (HF) configuration with the thin selective layer on the inside of the HF (see figure, left side). An improvement to these membranes would be to have the selective layer on the outside of the HF (right side). Such outer skinned PEMMs would have the advantage of an increased surface area per fibre and, allow fibres to be operated in the outside-in filtration configuration, enabling the use of new module geometries with superior mass transport properties {Futselaar et al, 1993} Moreover, it would enable the use of smaller mechanically stronger fibres, due to contaminants in the feed not being able to clog the bore of the fibres.

To prove that the concept of an outer skinned PEMM is feasible, we used the dry-jet wet spinning process to fabricate charged outer skinned HF support membranes and coated them through the layer-by-layer self-assembly process to form a PEMM. The results showed that we were able to form the first outer skinned HF PEMMs with a 3.5x increase in active membrane surface area to volume ratio as compared to commercial PEMMs. SEM images of the support fibres confirmed the existence of an asymmetric structure with an outer skin, and fluorescent imaging of the coated fibres confirmed that the polyelectrolyte multilayer was located on the outside of the fibre. The PEMMs showed nanofiltration properties ideal for removing OMPs with a pure water permeability of 3 L/(m2*h*bar) and a molecular weight cut-off of 300 Da. Our results thus clearly pave the way for the development of the next generation of HF nanofiltration membranes.

3:05 PM

Oral 45 - Tannic acid-coated hollow fiber for dehydration: From material to membrane and module

Lakshmeesha Upadhyaya (King Abdullah University of Science and Technology (KAUST))*

Around 40% of world energy in commercial and residential buildings is consumed by heating, ventilation, and air-conditioning units, and about 10% of greenhouse gases released to the environment are by cooling units. Conventionally, dehumidification of air condition units is based on vapor condensation and desiccant-based absorption. This process requires high capital and maintenance costs, the need for ample spaces, and, not the least, high energy utilization of up to 55% of the energy supplied to HVAC units. Membrane-based dehumidification is portrayed as a substitute due to its lower footprint and miniaturized prototype dimension, primarily when hollow fibers are employed.

In our research, we have developed a polyetherimide hollow fiber using our pilot-scale spinning line. These support fibers were coated with tannic acid, a naturally abundant polyphenol, and cross linked with hyperbranched polyethyleneimine that helps in cross-linking. The coated fibres exhibited water vapor permeance between 4200 GPU to 13,000 GPU with a selectivity of 133 to 47300. As a dehumidification unit for an HVAC system with 500 cfm of humid inlet air with 80-90% humidity at 33.5°C, about 6-9 m2 of membrane area is needed. An in-house centrifuge prepared the modules 30 cm long and 6 cm in diameter, carrying 500-700 fibers. The hollow fiber configuration with the advantage of miniaturization with higher packing density would give the final unit dimension of 0.28m x 0.2m x 0.2m, which is relatively small and portable. The Modules were tested in natural conditions for more than 380 days continuously. They showed a decline of 7% in water vapor transport with COP of 2.45, making it the most promising and efficient material. The material and prototype being proposed have potential applications in dehumidification for air conditioning and domestic as well as industrial dehydration processes. Its implementation could provide a sustainable solution with potential benefits for the community.

3:25 PM

Oral 46 - Developing a Thin Film Composite Facilitated Transport Hollow Fiber Membrane for Post Combustion CO2 capture

Oguz Karvan (EMI Twente BV)*; Ilse Pierik (EMI Twente BV); Jurjen Regenspurg (EMI Twente BV); Patrick de Wit (EMI Twente BV); Claudia Ruiz Ruiz (Aqualung Carbon Capture AS); Arun Kumar Panneer Selvam (Aqualung Carbon Capture AS); Paul A. Buchinger (Aqualung Carbon Capture AS); Michael Duthie (Aqualung Carbon Capture UK Limited); Saravanan Janakiram (Aqualung Carbon Capture AS)

Thin film composite (TFC) membranes are considered to be commercially attractive for post-combustion CO_2 capture, owing to the low cost and manufacturability of porous support when combined with a scalable selective layer coating. Such TFC membranes in hollow fiber morphology typically have three layers; a support with high surface porosity, an optional intermediate layer to prevent pore penetration and a permselective top layer. In this study, we showcase the development of a TFC membrane in collaboration with Aqualung Carbon Capture using their unique selective coating

Polysulfone was chosen to develop the highly porous support and silicon elastomer was chosen as the intermediate layer to prevent pore penetration of the dip coating solution of the permselective material. He/N_2 perm-selectivity and permporometry were used to characterize the porosity of the support. The porous support material has 12880 GPU CO_2 permeance and a He/N_2 selectivity of approximately 2.1. A commercially available silicon elastomer intermediate layer was batchwise applied by dip coating technique on single fibers by a speed controlled robotic arm. Fibers were subject to a series of washing steps and subsequently an intermediate layer was applied on pore filled fibers. Fibers with the intermediate (gutter) layer showed over 900 GPU CO_2 permeance and a CO_2/N_2 selectivity over 7.5. Current Investigation is focusing on increasing the permeance to over 2000 GPU while maintaining the selectivity.

Together with Aqualung Carbon Capture, a permselective layer comprising of a facilitated transport polymer has been coated on top of the silicone layer, to result in scalable hollow fiber TFC membrane for

CO_2 separation. To do this, secondary dip-coating step was applied resulting in fibers with over a 500 GPU CO_2 permeance with a mixed gas CO_2/N_2 separation factor of 30. Further optimization studies for the gutter and selective layer coating are ongoing.

Liquid Extraction and Contaminant Separation

1:45 PM

Oral 47 - XpressKleen 1 nm expanded PTFE membrane for harsh chemical filtration

Zhefei Li (Pall Corporation)*; SIAVASH DARVISHMANESH (Pall corporation); Catherine Agcaoili (Pall Corporation) Here we are reporting the development of a new generation of expanded polytetrafluoroethylene (ePTFE) membrane that boosts process reliability, reproducibility, and productivity in the semiconductor industry and is compatible with its various needs. According to the Semiconductor Industry Association, chip sales in 2024 will be more than US$500 billion as global demand continue to increase for electric vehicles, artificial intelligence, data centers, etc. Pall corporation ePTFE membrane technology products are designed to help chip manufacturers meet the increasing demands of single-wafer and batch-wet chemical processing. Pall's new ePTFE membrane is designed with improved surface kinetics (via the propriety process) for greater non-dewetting in aqueous and high-viscosity chemicals. While this membrane improved retention and de-wetting ability, there is no sacrifice to flow or service life compared to the previous generations of PTFE membranes in the market. This membrane brings a new level of contaminant removal capabilities to the wet etch chemical industry, going beyond current manufacturing barriers to remove the next level of contaminants (down to sub-2 nm in size and ppt level of concentration). The membrane's unique structure is designed to allow three zones of retention.

Pall's new PTFE membrane and unique surface chemistry enable it to operate at both high temperatures and harsh chemical filtration, like in Piranha filtration. The membrane also features improved cleanliness and significantly lowering the start-up time of the chemical process. The new PTFE membrane establishes new performance benchmarks in the marketplace. The membrane retention performance confirms being the tightest structure of the PTFE membrane in the market. It can significantly reduce defects during product manufacture while maintaining low differential pressures during dispensing or chemical processing.

2:05 PM

Oral 48 - 2,3-Butanediol Separations using Membrane Assisted Liquid-Liquid Extraction

Shailesh Singh Dangwal (Oak Ridge National Lab)*; Syed Islam (Oak Ridge National Laboratory); Blake Trusty (Oak Ridge National Laboratory); Zachary Coin (Oak Ridge National Lab); Ramesh Bhave (Ornl)

2-3 butanediol (BDO), an important component of fermentation broth having ~10wt.% composition, has varying application in the production of many important chemicals such as synthetic rubber, plasticizer, and as octane booster etc. Moreover, BDO also holds immense potential in its use as a component in the sustainable aviation fuel (SAF). In this work, we focused on the recovery of BDO from simulated fermentation broth using a continuous, energy-efficient, and scalable membrane assisted liquid-liquid extraction (MALLE) process. Different organic solvents such as hexanol and oleyl alcohol (OA) were used for the extraction of BDO from simulated fermentation broth. Using distribution coefficient studies, affinity of both organic solvents for BDO was first evaluated. Both hexanol and OA showed the ability to recover >90 wt.% BDO from the simulated fermentation broth. Additionally, low solubility of water in hexanol and OA also makes them a suitable candidate for recovering BDO from the fermentation broth. Organic solvents to aqueous feed ratio of 4 and 10 were able to recover >90% BDO from simulated fermentation broth using OA and hexanol, respectively. Presence of impurities (such as acetoin and glycerol) also did not impact the BDO extraction ability of the organic solvents. In a continuous MALLE system, BDO recoveries obtained from distribution coefficient studies were replicated for a ratio of 1, 2, 4, and 10 for hexanol as the organic solvent. MALLE offers multiple advantages over traditional solvent extraction such as it is more energy efficient, single-step continuous process, scalable, and low-cost process. Additionally, MALLE prevents the emulsion formation which is a key advantage over traditional solvent extraction process. Thus, MALLE holds immense potential as a continuous, scalable, cost efficient and energy efficient technology for the recovery of BDO.

2:25 PM

Oral 49 - Continuous membrane demulsification for in situ solvent extraction of carboxylic acid

Yian Chen (National Renewable Energy Laboratory)*; Eric Tan (National Renewable Energy Laboratory); Jacob Dempsey (National Renewable Energy Laboratory); Patrick Saboe (National Renewable Energy Laboratory); Violeta Sànchez i Nogué (National Renewable Energy Laboratory); Jeffrey Linger (National Renewable Energy Laboratory); Gregg Beckham (National Renewable Energy Laboratory)

Solvent-based liquid-liquid extraction (LLE) is a common in situ separations strategy for many biochemicals produced through fermentation but is challenged by emulsion formation. In this study, we aim to show that a hydrophobic polytetrafluoroethylene (PTFE) membrane-based emulsion separator (MBES) enables continuous LLE for the recovery of an exemplary fermentation product, butyric acid. MBES-assisted LLE starts with sufficient mixing of the two phases to promote emulsion formation. Emulsification ensures ample contact between the two phases through increased effective interfacial area, and thus enhances the partitioning rate of a target product into the extraction solvent. After achieving equilibrium, the MBES system then continuously separates the emulsion, by leveraging surface tension and fine-tuning pressure differential, into a product-rich organic solvent and product-lean aqueous phase. At an equivalent butyric acid throughput, process modeling estimates that integrating MBES-assisted LLE can enable a 66% and 40% reduction in process costs (the sum of CAPEX and OPEX) and greenhouse gas emissions, respectively, compared to the traditional downstream bioprocessing. Although membrane protein fouling led to reduced membrane flux and phase separation efficiency, common fouling mitigation strategies including the incorporation of ultrafiltration and periodic membrane backwashing effectively recovered the membrane performance. Overall, MBES systems can potentially enable continuous LLE processes in bioprocessing separations, including where emulsion formation is a challenge.

2:45 PM

Oral 50 - Membrane nanoprecipitation enables the continuous production of uniform nanoparticles at higher concentrations of the polymer solution than with a batch system

Lidietta GIORNO (National Research Council of Italy - Institute on Membrane Technology (CNR-ITM))*; Antonio Condello (National Research Council of Italy - Institute on Membrane Technology (CNR-ITM)); Emma Piacentini (National Research Council of Italy, Insitute on membrane Technology, CNR-ITM)

Nanoprecipitation is a well-known process that is used on a small batch scale to obtain spherical particles with a uniform size distribution on a nanoscale. Porous membranes can be operated as continuous micromixers with high throughput and enable continuous nanoprecipitation.

The lecture illustrates a comparison between a conventional batch nanoprecipitation system and a membrane nanoprecipitation in producing nanoparticles made of synthetic polymer or protein. A systematic study on the influence of phase composition, solution concentration, and solvent/non-solvent volumetric ratio allowed us to identify the zone of the three-phase diagram where non-solvent-induced phase separation promotes self-assembly of the polymer into stable nanoparticles with uniform size distribution (the so-called stable Ouzo zone). For both types of nanoparticles, we found that membrane nanoprecipitation extends the stable Ouzo region compared to batch nanoprecipitation. In other words: When the solution was forced to flow through a porous membrane and added to the non-solvent flowing along the other side of a membrane, nanoparticles with uniform size distribution could be obtained even at high solute concentrations, where the batch process results in nanoparticles with non-uniform size distribution. We also found that nanoparticles with the same size and size distribution were obtained using different pore sizes (from submicron to micron range) and that the most important parameter affecting the size and size distribution of nanoparticles is the ratio between the flow rate of solvent and non-solvent.

Membrane nanoprecipitation also resulted in uniform nanoparticles when the non-solvent was forced through the pores and added to the polymer solution, which can prevent fouling.

Acknowledgments

Authors thank the Project Nutrage funded by CNR project FOE-2021 DBA.AD005.225.

3:05 PM

Oral 51 - High temperature ceramic membrane separations for carbon circularity: From lube oil purification to chemical recycling of plastics

Pieter Vandezande (VITO)*; Anita Buekenhoudt (VITO); Efecan Pakaner (VITO); Fady Nahra (VITO); Pedro Borges (VITO); Mohammed Nazeer Khan (VITO); Mohammed El Ibrahimi (VITO); Miet Van Dael (VITO)

For decades, mankind has been living in a 'make, take, dispose' economic model. Plastics, with their unrivaled properties and low cost, are an exponent of this way of living. While global plastics production reached 400 million tons in 2022, it is estimated that today nearly 50% of the plastic waste in the EU is landfilled, roughly 25% incinerated and only 25% recycled. Another example are used lubricating oils (ULOs), equally derived from fossil feedstocks and essential in a wide range of industrial applications, but today recycled only to a limited extent and not in the most energyefficient way. To mitigate greenhouse gas emissions, recent legislation calls for far higher recovery and recycling rates of these materials at the end of their service life. Thanks to their inherently energy-efficient, clean and mild character, as well as flexibility and scalability, membrane processes are fully aligned with current industry drivers towards electrification, carbon circularity and minimal discharge. Therefore, membranes are considered a powerful tool, even key enabling technology, for tackling tomorrow's separation challenges in chemical related industries. Recovery of base lubricating oils from ULOs and of virgin polymers from solvent-based plastics mixtures, targeting removal of critical impurities such as lubricant and plastics additives, calls for robust membranes that can cope with highly viscous oils and solvent streams at high temperatures. To overcome these stability challenges, tubular ceramic membranes with tailored surface chemistries are being studied. This research fits in the ongoing EU project CUMERI and Belgian project RENOVATE, focused on separation development for ULO purification and dissolution-based plastics recycling, respectively. Separation of such challenging hydrocarbons streams using intensified membrane processes will contribute to increasing recycling rates, enhancing carbon circularity and reducing the CO2 footprint of the chemical industry.

3:25 PM

Oral 52 - Targeted Separation Scheme of Polyurethane Depolymerization Products

Taysha B Telenar (Arizona State University)*; Timothy Long (Arizona State University); Matthew Green (Arizona State University)

In the United States, end of life polyurethane foams such as mattresses or cushions go predominantly to landfills. Mechanical recycling can prolong the inevitable fate of these materials, but not indefinitely. An emerging option is chemical recycling: breaking down these foams into smaller re-polymerizable feedstocks with a goal of creating a polyurethane circular economy. Chemical recycling creates a chemical mixture, requiring separation mechanisms to produce contaminant-free feedstocks. The proposed separation methods utilize membrane technology through a variety of processes such as pervaporation and filtration because these processes require lower energy. Pervaporation is a chemical potential driven process which uses dense polydimethylsiloxane (PDMS) membranes to remove any volatile components such as methanol, ethanol, butanediol, and ethylene glycol early in the process. Diffusion experiments show that all volatile components mentioned have diffusion coefficients on the same magnitude as ethanol, which is industrially separated through pervaporation. Pervaporation experiments provide permeability of each volatile component. Sorption coefficients represent the thermodynamic interactions as particles move through the membrane and are dependent on the diffusion coefficients and permeability. The proposed microfiltration process is pressure driven that uses porous bisphenol A polysulfone (BPA-PS) and 4,4′-dihydroxydiphenylmethane polysulfone (BPF-PS) membranes to remove larger components or oligomers such as polypropylene glycol (PPG). Microfiltration membrane process conditions are integral to the overall performance of membranes. Polymer composition, molecular weight, concentration, thickness, and operating pressures influence water permeance through the membrane. Higher molecular weight (~40 kDa) polymers at 15 wt% loading in N-methyl pyrrolidone (NMP) show the best performance for pure water.

Abstracts – Oral Presentations, Monday Evening

Presenting authors are indicated by an asterisk*.

Desalination and Osmotic Processes I

4:00 PM

Oral 53 - Compaction of high-pressure water treatment membranes: real-time quantification and support layer contribution

Weijian Ding (University of Southern California)*; Kexin Ma (University of Southern California); Amy Childress (University of Southern California)

Water treatment membranes play crucial roles in applications such as seawater desalination, wastewater treatment, potable water reuse, and resource recovery. Their performance, including transmembrane water flux and salt rejection, can be adversely affected by compaction. Distinguishing between membrane compaction and other factors (e.g., membrane scaling, fouling, or reduced surface tension) affecting performance degradation, especially during long-term higher-pressure operations, is challenging. In a prior study, we introduced a novel method, combining electrical impedance spectroscopy with dynamic mechanical analysis, to quantify single-layer homogenous membrane compaction under low pressure (up to 12.5 psi). In this study, we extend the method's capabilities to quantify real-time compaction of multi-layer heterogenous nanofiltration and reverse osmosis membranes under higher pressures, reaching up to 330 psi. By measuring membrane instantaneous compaction, creep and recovery, and hysteresis, we quantify the immediate and longer-term response (fatigue) of the material. Unexpectedly, our findings reveal that air pockets between the membrane layers contribute more to total compaction than compaction of the support layers themselves, contrary to previous studies that generally neglected the existence of air pockets and their contribution to membrane compaction. We introduce the concept of "compaction thresholds" to elucidate how individual membrane layers, including the polyamide selective layer and underlying support layers, contribute to total compaction. Finally, we compare the measurement results obtained from our method with the membrane film test, a common representation of membrane compaction in homogenous membranes. Contrary to prior findings, our research reveals that this test inaccurately describes the compaction properties of both homogenous (e.g., single-layer membrane distillation membranes) and heterogeneous (e.g., multi-layer NF and RO) membranes.

4:20 PM

Oral 54 - Membrane heterogeneity alters our interpretation of effective energy barriers to transport

Nathanael S Schwindt (University of Colorado Boulder)*; Mor Avidar (Technion); Razi Epsztein (Technion - Israel Institute of Technology); Anthony Straub (University of Colorado Boulder); Michael Shirts (University of Colorado Boulder)

Major efforts in recent years have been directed towards understanding the molecular determinants of transport in reverse osmosis and nanofiltration membranes. Transition-state theory is an increasingly common approach to explore mechanisms of transmembrane permeation with molecular details. This theoretical framework relates molecular jumps to free energy barriers to membrane transport. However, most investigations treat all these free energy barriers in the membrane as equal, which neglects the inherent structural and chemical heterogeneity in polymeric membranes. In this study, we expand the transition-state theory framework to include distributions of free energy barriers in order to incorporate more realistic membrane heterogeneity. We develop a novel expression relating individual, molecular barriers to experimentally observable free energy barriers. We show that the highest free energy barriers along the most permeable paths, rather than typical paths, provide the largest contributions to the observable free energy barrier. We identify and analyze multiple levels of heterogeneity that must be considered when designing separations membranes from the molecular level. Experimentally observed energy barriers must be interpreted in terms of the underlying molecular barriers, and naive interpretations of these effective barriers can lead to incorrect assumptions about the transport mechanisms within polymeric membranes.

4:40 PM

Oral 55 - Evidence for solution-diffusion in pressure-induced transport of solvent through dense polymer membranes

Kevin Reimund (University of Texas at Austin)*; Rahul Sujanani (The University of Texas at Austin); Jorge Hernandez (The University of Texas at Austin); Kristofer Gleason (The University of Texas at Austin); Manish Kumar (University of Texas at Austin); Benny Freeman (The University of Texas at Austin)

Understanding the molecular mechanism of liquid transport through polymeric membranes has been a topic of debate since the development of the first reverse osmosis membranes. Recently, a number of authors have suggested that the solution-diffusion model is incapable of accurately describing membrane transport, particularly in reverse osmosis, and that its theoretical underpinnings, namely the assumption that the membrane pressure is both constant and equal to that of the feed solution, may be fundamentally flawed. Mechanical models, based on the concept of pressure drop through a porous network of subnanometer pores, have been proposed as an alternative hypothesis. However, we investigate the transport of solvents across dense polymer membranes at transmembrane pressures up to 240 bar and observe behavior consistent with the solution-diffusion model, including so-called limiting flux behavior for water. We correlate this behavior with measured solvent sorption isotherms by utilizing the relationship between solvent activity and pressure proposed by the solution-diffusion model. In conjunction with established models for predicting penetrant diffusion in polymers, the relationship between pressure and flux can be predicted with knowledge of the sorption behavior and diffusivity. In some cases, a quantitative prediction can be made without the use of any adjustable parameters, suggesting the applicability of the assumptions underlying the diffusion model as well as the mechanical assumption of constant membrane pressure which enables computing the chemical potential gradient across the membrane. To corroborate this, we compute the pressure profile at the interface of idealized dense membranes and rigid porous supports to demonstrate that pressure fluctuations caused by this contact decay rapidly with distance away from the interface. These results suggest that assuming a constant membrane pressure is reasonable for many membrane systems.

5:00 PM

Oral 56 - Is the Role of Concentration Polarization Overstressed in Membrane Filtration Processes?

Lianfa Song (Texas Tech University)*

Concentration polarization is regarded as a major constrain on the performance of the membrane processes. Hence, a primary measure for performance improvement is to minimize concentration polarization in the design and operation of the membrane processes. Though concentration polarization is one of the most studied topics in membrane filtration processes, current understanding of this phenomenon is still circumscribed within the theoretical framework that was developed in the early stage of this technology. It is commonly accepted that the framework was constructed basically on the oversimplified cases or questionable assumptions. There are always concerns if the role of concentration polarization in the membrane filtration processes can be well assessed within this theoretical framework.

The author will show that the impact of concentration polarization on the performance of membrane filtration in many cases were unduly overstressed with the problematic modeling methods and inappropriate conceptualizations. The traditional way to model concentration polarization overly emphasizes the impact of the hydraulic conditions. Actually, a more fundamental factor for concentration polarization is the accumulation of the "retained" species be the membrane. More reasonable modeling method and appropriate assessment of the role of concentration polarization in membrane processes will be presented.

5:20 PM

Oral 57 - Modeling and experimental analysis of pH influence in boron separation in reverse osmosis

Fernan David Martinez Jimenez (KAUST)*

Interactions between ions and reverse osmosis (RO) membranes, mainly concerning pH, are crucial for accurate modeling of boron rejection. RO membranes have an active layer (AL), usually made of polyamide, responsible for ion separation. The AL contains charged functional groups that reject similarly charged ions. Charge and concentration of the functional ions depend on the pH; hence, changes in pH affect the ion rejection.

Studies have shown the relationship between feedwater pH and boron rejection by different membrane types1–3. However, they considered the carbonate system in the water. Carbonate ions interact with boron speciation by changing the solution pH and by electromigration, thus affecting boron transport through the membrane and the rejection behavior. These studies also presented a restricted pH range and limited measurements. Therefore, a more comprehensive study is needed to elucidate the variations in boron rejection at different feedwater pH, considering different salinities (Na+ and Cl- concentrations) and the influence of the carbonate system. In this context, we investigated boron removal using the brackish and seawater membranes across a broad pH range and varying salinities. We also developed a solution-friction model to explain the boron removal process.

Boron rejection was influenced by carbonates, especially in the pH range of 6 to 8, when the measured rejection was higher in the presence of carbonates. This effect was not previously reported, highlighting the importance of carbonates in boron rejection studies. The simulations effectively represented permeate pH, Na+ and boron rejection at low salinity. In addition, the prediction of boron rejection needs to be improved for pH < 6 and in the presence of carbonates. Finally, simulations and experimental data comparisons revealed three different permeate pH equilibria: after crossing the membrane, in the permeate pipes, and the samples in contact with the atmosphere.

5:40 PM

Oral 58 - Flexible RO-NF System Configuration Operation in the Mode of Partial Concentrate Recycle for Inland Water Desalination

Yakubu Jarma (University of California, Los Angeles)*; Beau Akhavain (University of California, Los Angeles); Yoram Cohen (Professor)

Reverse osmosis (RO) and nanofiltration (NF) technologies are well established for brackish inland water desalination. However, high salinity inland water desalting requires elevated applied feed pressure which can significantly raise the system's capital cost. Moreover, concentrate ("brine") management and membrane fouling/mineral scaling remain as major impediments to wider scale adoption of inland water desalination. Thus, high recovery desalting is desired for brine volume minimization. However, high recovery demands an expanded system footprint (i.e., greater number of membrane elements) and high-pressure pumps both increasing the system capital cost. To address the above challenges, a novel flexible RO-NF (FLERONF) desalination system configuration was developed allowing operation in the mode of partial concentrate recycle. Studies were conducted with a pilot FLERONF (0.2-1 cubic meter/h capacity) in various RO-NF configurations. Feed water was of salinity range of ~3,000-30,000 mg/L total dissolved solids (TDS) and high calcium sulfate and calcium carbonate scaling potential. FLERONF, which accommodated up to six 2.5-inch x 40-inch spiral-wound membrane elements in series, had feed a filtration pretreatment train, antiscalant dosing, and advanced system control with real-time monitoring and data acquisition. Through optimal NF/RO elements configuration the applied feed pressure was significantly reduced relative to conventional RO configuration. Moreover, FLERONF operation was with lower specific energy consumption (SEC), which was also aided by increasing the per-pass recovery. Also, reduced system footprint and high recovery was enabled via concentrate recycle operation. Membrane scaling was avoided via implementation of optimal periodic permeate flush (along with antiscalant dosing). The produced permeate quality was suitable for agricultural irrigation and potable use, although second-pass recovery was necessary for feedwater at the high salinity range.

Novel Synthetic Strategies for Gas Separation

4:00 PM

Oral 59 - Permeance Prize: Utilizing Molecular Design to Integrate Ionic Liquids within High Performance Polymer Gas Separation Membranes: Unprecedented Opportunities for Control over Material Properties and Performance

Jason Bara (University of Alabama)*

This talk will discuss the role that ionic groups and ionic liquids (ILs) can play in developing the next generations of high performance gas separation membranes by reconsidering the molecular design of ionic polymers (e.g., poly(ILs) and ionenes). The unique opportunities are afforded through imidazole-containing aromatic diamine or diol monomers which are compatible with, and complementary to, virtually any polymer backbone including polyimides, polyamide, Troger's Base, polyesters, and polyethers, thus allowing ionic groups and ILs to be introduced directly into state-of-the-art membrane architectures. This approach allows for the influence of polymer structure on material physical properties and gas separation performance to be elucidated systematically so as to develop optimized membrane materials and address challenges such as plasticization and aging.

4:20 PM

Oral 60 - Few-Cycle Atomic Layer Deposition to Nanoengineer Polymeric Materials for Efficient Gas Separations

Leiqing Hu (University at Buffalo)*; Won Il Lee (Stony Brook University); Ashwanth Subramanian (Stony Brook University); Erda Deng (State University of New York at Buffalo); Kim Kisslinger (Brookhaven National Laboratory); Shouhong Fan (University of Colorado); Vinh Bui (University at Buffalo); Yifu Ding (Membrane Science, Engineering, and Technology (MAST) Center, Department of Mechanical Engineering, University of Colorado Boulder); Chang-Yong Nam (Stony Brook University); Haiqing Lin (The State University of New York at Buffalo)

Atomic layer deposition (ALD) creates uniform sub-nanometer films on a variety of surfaces and nanopore walls and has been used to modify polymers to improve surface affinity towards specific molecules, solvent resistance, and barrier properties to gases and vapors. Here, we demonstrate that few-cycle ALD can be used to engineer polymer-based membranes to enhance molecular sieving capability for gas separations. First, polybenzimidazole (PBI) is exposed to trimethylaluminum (TMA) and water vapor alternatively, forming AlOx at a sub-nanometer scale. The treated PBI exhibits improved molecular size-sieving ability and counterintuitively, gas permeability. Specifically, the ALD not only deposited an AlOx layer on the surface but also enabled the TMA to infiltrate and react with the bulk PBI to form an AlOx network, disrupting polymer chain packing and increasing chain rigidity. For instance, 1-cycle ALD treatment remarkably increased H2 permeability by 120% - 270% and H2/CO2 selectivity by 30% at 35 - 200 oC. Second, we also observed that the microcavities of polymer-derived carbon molecular sieve (CMS) materials can be effectively tuned using a modified ALD process. The AlOx atomically deposited onto the micropore surface of CMS, leading to decreased porosity and pore size and thus enhanced size-sieving ability. For example, 1-cycle ALD treatment increased H2/CO2 selectivity from 4.0 to 33 at 100 oC for a PBI carbonized at 600 oC. This approach was also successfully applied for polyimide-derived CMS materials to enhance their size-sieving ability, demonstrating the versatility of the technique. To conclude, this study showcases a facile and scalable way of engineering polymeric and CMS membranes at a sub-nanometer level to improve molecular separation performance.

4:40 PM

Oral 61 - Amine Modified PIM-1 for Highly Selective Olefin/Paraffin Separation

Bo Wei Cynthia Chen (University of Notre Dame)*; Casey O'Brien (University of Notre Dame)

Cryogenic distillation is a common practice in the current refinery industry to purify light hydrocarbon gases from crude oil streams, but it is a thermally intensive process with overwhelming consumption of energy. In order to lower the carbon footprint associated with industrial distillation processes, literature has shown that membrane separation processes can reduce energy consumption up to 90%. However, membranes have limitations on their gas separation efficiency, as it is dependent on the intrinsic properties of the material. The gas diffusivity and solubility in a polymeric membrane can be fine-tuned by designing the structure of the polymer. Polymer of Intrinsic Microporosity (PIM-1) is well known for its bulky, ladder-shaped backbone that creates high free volume, leading to high gas diffusivity, but suffers from a low gas selectivity. By introducing amine groups into the PIM-1 backbone, the hydrogen bonding forces creates tighter packing in the membrane, creating new gas separation characteristics and efficiency. This study proposes to develop a novel membrane separation system to selectively separate hydrocarbon gases by using amine functional groups. Initial investigation of amine-modified PIM-1 (PIM-NH2) shows higher propylene/propane separation performance than PIM-1. At the same time, PIM-NH2 has lower propylene permeability and faster aging compared to PIM-1. The project explores the gas separation efficiency of different amine composition and future plans for aging mitigation.

5:00 PM

Oral 62 - A New Microcapillary Film Carbon Molecular Sieve Membrane

Jay (Junqiang) Liu (The Dow Chemical Company)*

Polymeric membranes, in the form of plate-and-frame, spiral wound, and hollow fiber, have been commercialized for light gas separations (i.e., O2/N2, H2/N2, CO2/CH4) since the 1980s. These membranes are not suitable for large scale hydrocarbon gas separations (i.e., olefin/paraffin) because of the membrane plasticization that results in lower permeance and limited selectivity. Carbon Molecular Sieve (CMS) membrane is considered as a promising next generation membrane technology, which offers both improved separation performance and stability in hydrocarbon stream. Several companies have tried to scale up the CMS membrane technology, but none have succeeded yet. One of the biggest challenges is the handling of the fragile carbon material. Membrane breakage can happen during module making and usage, which prevents the scale-up production.

For the first time, we propose a new microcapillary film CMS membrane to mitigate the critical issue of scalability. Microcapillary film has good mechanical strength like a closely packed bundle of fibers. It also has large effective surface area because permeation takes place from both sides of the membrane to the internal microcapillaries. A prototype 0.5" wide microcapillary CMS film with 42 microcapillary channels was prepared to demonstrate the concept. First, polyvinylidene chloride copolymer was melt extruded with a custom-made die with insert-pins that can create 42 microcapillaries by the air flow through them. The obtained microcapillary film was then pyrolyzed using a two-step process to 600ºC under inert purge. The final microcapillary CMS film thickness is about 230 micron and each of the 42 microcapillary diameter is about 80 micron. The microcapillary CMS film showed good separation properties in mixed gas olefin/paraffin tests at 52psig and 35-100ºC.

5:20 PM

Oral 63 - The Effect of Crosslinker Concentration on Competitive Release Kinetics and Physical Characteristics of Thermo-Responsive, Lignin-Based Soft Composites

Missoury Wolff (Clemson University)*; Eric Davis (Clemson University)

The use of lignin in soft composites has gained recent attention due to its various favorable qualities, such as biocompatibility, antimicrobial and antibacterial properties, and sustainability of sourcing. Lignin, an abundant phenolic biopolymer, can be used in conjunction with stimuli-responsive materials, such as poly(N-isopropylacrylamide) (PNIPAm), as these materials are able to undergo a volume change at a temperature close to that of human physiological conditions. The fabrication of interpenetrating networks (IPNs) containing PNIPAm and other crosslinked polymers have been explored to achieve this volume transition. However, traditional methods investigate the role of exclusively petroleum-based materials in the fabrication of these soft composites. In this study, we fabricated soft composites containing: (1) a thermo-responsive polymer, PNIPAm; (2) a hydrophilic polymer, poly(vinyl alcohol) (PVA); and (3) a sustainable biopolymer, lignin. To form the IPN of the hydrogel, the concentrations of crosslinkers glutaraldehyde and N,N'-methylenebisacrylamide were varied between 5 and 15 mass percent of their respective polymers. After fabrication, the mechanical properties and drug-release kinetics of the soft composites were characterized at both room temperature (~21 °C) and 40 °C. Specifically, physical properties were measured using dynamic mechanical analysis and mechanical indentation. In addition, the equilibrium water uptake was measured. Finally, the diffusion of caffeine through the network was analyzed via ultraviolet-visible spectroscopy. It was observed that the addition of lignin to the soft composites altered all properties measured. In particular, the diffusion of caffeine was suppressed with the addition of lignin, where complete release of each drug was seen to be significantly slower in membranes containing lignin. Furthermore, the crosslinker concentration was also seen to impact the release kinetics of these soft composites.

5:40 PM

Oral 64 -

Tailored Membrane Formation

4:00 PM

Oral 65 - Sulfonated Polyamide Thin Film Composite Membranes for Electrodialysis

Deniz Ipekci (University of Connecticut)*; Yasmin Teixeira Bimbatti (UCONN); Xingyu Wang (University of Connecticut); Inara Oliveira (University of Virginia); Anna Harris (University of Virginia); Geoffrey M Geise (University of Virginia); Baikun Li (University of Connecticut); Jeffrey R McCutcheon (University of Connecticut)

Electro-membrane processes, such as electrodialysis (ED), are finding important niche applications in desalination, water treatment, and resource recovery. Customizing ion exchange membranes (IEMs) is critical to tailor their use for these variety applications. Simultaneously, IEMs on the market today tend to be prohibitively expensive and lack the necessary selectivity to achieve the desired goals of generating clean water or capturing specific metals.

Numerous ionomers have been developed for ion exchange membranes. These monomers are all processed in the same approach (typically through extrusion or casting). The resulting membranes are thick, expensive, and have high ohmic resistances. In this work, we propose the use of a thin film composite (TFC) design for IEMs in hopes of reducing their cost and resistance. We propose the use of a new manufacturing approach, electrospray deposition, to synthesize these highly selective and low-resistance membranes in a scalable and inexpensive way.

We have selected the aromatic polyamide chemistry (typical with reverse osmosis (RO) membranes) as a platform material for making these TFC membranes. A common material for making TFC membranes, polyamides exhibit some ion exchange capacity due to the presence of carboxylic acid groups in the non-crosslinked part of the polyamide. We enhance the ion exchange capacity of the polyamide by introducing sulfonate groups through the use of sulfonated co-monomers (such as the sulfonated analog of MPD) during the synthesis process. This incorporation is specifically enabled through the electrospray process since conventionally formed films through interfacial polymerization are unlikely to incorporate bulkier and less reactive sulfonated monomers into the polymer matrix.

4:20 PM

Oral 66 - Compaction-resistant Thin Film Crosslinked Composite (TFX) Ultra-high Pressure Reverse Osmosis

Jishan Wu (UCLA)*; Eric Hoek (UCLA)

This research explores the capability of ultra-high pressure reverse osmosis (UHPRO) to supplant energy-intensive thermal desalination, offering potential cuts in energy usage and brine concentration costs by up to 50%. Notwithstanding, the performance of commercial RO membranes significantly degrades under ultra-high pressure due to intense membrane compaction and embossing. We thoroughly examine the effects and mechanisms of this composite membrane deformation at both active and support layers. We also introduce a groundbreaking RO membrane design with exceptional compaction resistance. These newly developed thin film crosslinked composite (TFX) membranes demonstrate remarkable compaction resilience under ultra-high pressures - up to 200 bar - thereby positioning RO membrane-based brine concentration as a contender to thermal methods. Unlike traditional RO membranes, which lose up to 45% water permeance at 200 bar due to physical compaction, these innovative membranes retain 85-95% of their original water permeance at the same pressure. Moreover, the novel membrane exhibits a thickness reduction of less than 15%, in stark contrast to the over 40% thickness reduction seen in conventional composite membranes. Beyond their impressive compaction resistance, the best performing novel membranes deliver salt rejections surpassing 99% for NaCl solutions with concentrations of up to 200,000 mg/L.

4:40 PM

Oral 67 - CFD modeling of high-flux plate-and-frame membrane modules for industrial carbon capture

Cheick Dosso (Carnegie Mellon University)*; Hector A Pedrozo (Carnegie Mellon University); Thien Tran (National Energy Technology Laboratory); Lingxiang Zhu (National Energy Technology Laboratory); Victor A Kusuma (NETL); David Hopkinson (NETL); Lorenz Biegler (Carnegie Mellon University); Grigorios Panagakos (Carnegie Mellon University)

Industrial carbon dioxide (CO2) emissions represent an important share of the global emissions. In the pursuit of short-term solutions to mitigate atmospheric CO2 levels and strive to achieve a net-zero scenario by 2050, carbon capture technologies (CCT) emerge as promising strategies. Looking for efficient CCT, membrane process has been emerging as an energy-efficient technology for CO2 capture in the past two decades.

In this work, we explore the use of CO2-selective flat sheet membranes for capturing CO2 from point sources. We employ Computational Fluid Dynamics (CFD) models to design high-flux plate-and-frame membrane modules to achieve uniform flow distribution among membrane elements, minimize dead-end zones, and ease the common concentration polarization issue for gas separation membranes. The goal is to drive membrane technology improvements by providing better module designs for given membrane properties and operating conditions.

Our modeling approach considers multiphysics processes, including fluid flow and transport of concentrated species within the membrane module. To validate the model, experiments are collected using lab-scale 3D-printed plate-and-frame modules under various conditions, utilizing a flue gas with 30% CO2 (molar basis) at flowrates ranging from 100 to 600 sccm. The membrane properties include a CO2/N2 selectivity of 28 and a CO2 permeance of 1600 GPU. The feed side is kept at a pressure of 18 psia, the permeate side at 2.9 psia, and the temperature is set at 22 degrees Celsius.

Numerical results show good agreement between experiments and model predictions for CO2 recovery, CO2 mole fraction in the retentate and permeate, and stage-cut. The CFD model elucidates the effect of hydrodynamics on CO2 mass transfer and facilitates extraction of the CO2 concentration profiles. The combination of CFD and experiments enhances our understanding of the process, paving the way to design efficient CO2 separation based on membrane systems.

5:00 PM

Oral 68 - Free Energy Surface Landscape of Interfacial Polymerization for Polyamide Membrane Formation

Cheick Dosso (Carnegie Mellon University)*; Hector A Pedrozo (Carnegie Mellon University); Thien Tran (National Energy Technology Laboratory); Lingxiang Zhu (National Energy Technology Laboratory); Victor A Kusuma (NETL); David Hopkinson (NETL); Lorenz Biegler (Carnegie Mellon University); Grigorios Panagakos (Carnegie Mellon University)

Conventional thin film composite polyamide membranes formed by interfacial polymerization (IP) has been widely used for water purification with nanofiltration (NF) and reverse osmosis (RO) applications. More recently, polyamide membranes have been developed for organic solvent resistant NF membranes. IP employs fast irreversible polymerization at the interface between the multifunctional monomers, m-phenylene diamine (mPD) in aqueous phase and trimesoyl chloride (TMC) in the organic phase yielding thick or thin films with low or high water permeance and good solute rejection. It is difficult to control the reaction leading a rough surface. Molecular layer-by-layer (mLBL) deposition has the potential to generate a kinetically controlled smooth membrane with desired pore structure and porosity. In order to better understand the reaction between mPD and TMC during IP, it is desirable to determine the reaction free energy and reaction rate as well as the effects of solvent, additives on the polymerization reactions. Here ab initio molecular dynamics (MD) simulations coupled with metadynamics (MTD) were performed to determine the free energy landscape for the IP reaction under different solvent conditions. Our results indicate that solvent plays a major role on the activation energy of the reaction and could potentially speed up or slow down the reaction for better control of the polyamide membrane structures.

5:20 PM

Oral 69 - The influence of additives on polystyrene membranes created by temperature-induced phase separation

Sulaiman Dhameri (University of Kentucky)*; Ibrahim Imam (University of Kentucky); Usman Abbas (University of Kentucky); Qing Shao (University of Kentucky); Gosia Chwatko (University of Kentucky)

Temperature-induced phase separation (TIPS) is one of the standard methods to create membranes in academia and industry. The upper critical solution temperature (UCST), resulting from intermolecular interactions in polymer-solvent systems, can be crucial in inducing a phase separation. At lower temperatures, these interactions form polymer-rich and polymer-poor phases, serving as the matrix and pores of the membrane, respectively. Current research on sustainable UCST-based membrane fabrication has been limited, particularly in achieving phase transitions at lower temperatures. This study explored the UCST of polystyrene-terpineol solution for synthesizing membranes. In particular, we investigated the influence of additives, oleic acid, gamma-valerolactone, and limonene on UCST and membrane performance. The cloud point temperatures were determined visually. Scanning electron microscopy was employed to assess pore size and pore shape, while porosity was measured using the gravimetric method. Water flux and bovine serum albumin rejection were tested through dead-end filtration. Our findings highlight the significant impact of the hydrogen bonding on UCST, with an observed change of over 35°C when varying additive concentrations from 0-15 wt%. Additive-free membranes and membranes containing 5 wt% of additives exhibited the average pore size of the top surface less than 50 nm, rendering them suitable for ultrafiltration applications. In conclusion, our study presents a method to manipulate UCST by incorporating small-molecule additives with diverse hydrogen bond propensities.

5:40 PM

Oral 70 - The dynamics of Interfacial Polymerization: insights on film stability during formation

Guy Z Ramon (Technion - Israel Institute of Technology)*; Adi Ben Zvi (Technion)

Thin-film polyamide (PA) selective layers fabricated via interfacial polymerization (IP) form the core of membrane-based water purification. Despite widespread commercial application, the exact details of reaction dynamics and film formation are not fully understood. Unravelling this fundamental knowledge is critical for controlling the IP reaction and designing better membranes for different applications. However, studies mostly rely on post-fabrication characterization, due to the challenging nature of IP that makes it almost impossible to monitor. A particular aspect of IP is the question of the resultant film morphoogy, its origin and impact on resultant membrane properties. Here, we study reaction dynamics and film formation via microscopy-based particle tracking, performed during IP, with reaction parameters systematically varied. Fluorescent particles located in the aqueous phase are observed by confocal microscopy during an IP reaction performed in a microfluidic device. Particle trajectories and motion parameters are calculated, determining the nature of aqueous-phase motion, used as an indication of interfacially-driven hydrodynamic instabilities. The underlying assumption here is that IP can induce an interfacial instability that alters film morphology, where a transition from stable to unstable state corresponds to a morphological shift of the resultant film, from smooth to crumpled. Resultant film morphologies are characterized using SEM, and AFM. This presented methodology allows us to close the loop in the fundamental understanding of synthesis-morphology relationship in PA membranes, which is an important step towards controlling reaction conditions to fabricate better performing membranes.

Lithium Separation

4:00 PM

Oral 71 - Co-Ion Specific Effect on Li Salt Sorption in Cation Exchange Membrane

Rachel Z Huang (UT Austin)*; Lynn Katz (The University of Texas at Austin); Benny Freeman (The University of Texas at Austin)

The demand for lithium (Li) has surged due to the widespread adoption of Li-ion batteries and the ongoing electrification of the grid. Electrodialysis devices show promise in efficiently extracting and concentrating Li^+ in aqueous solutions, where Li^+ is transported through a CEM under the influence of an electrochemical potential gradient. Recent studies have revealed that the sorption of salts in CEMs

can be influenced by the co-ion (anion of the salt) charge density. For example, in Na_2SO_4, SO_4 (-2), unlike Cl (-1), can form ion pairs and be absorbed into the membrane as a reduced-charge species $NaSO_4$ (-1). The ion-pairing effect reduces the Donnan-exclusion effect on the co-ion.
For salts with the same cation, the ability to form ion pairs increases with the anion charge density. In this study, we investigated Li salt sorption in a standard CEM (CR 61) by systematically varying the co-ion charge density. The co-ions, ranging roughly from low to high charge density, are bistriflimide (TFSI (-1)), triflate (OTF (-1)), nitrate (NO_3 (-1)), chloride (Cl (-1)), sulfate (SO_4 (-2)), and oxalate (C_2O_4 (-2)). These ions are of interest in a wide range of energy, water, and resource recovery applications.
We found that the salt concentration absorbed into the membrane is lowest at a medium anion charge density. Additionally, we employed a modified Donnan-Manning model that accounted for ion pairing to calculate membrane salt concentration and found reasonable agreement with the experimental data.

4:20 PM

Oral 72 - Mixed counter-ion sorption in ion-exchange membranes: understanding the role of membrane water content

Harsh Patel (University of Michigan)*; José Carlos Díaz (University of Michigan); Jovan Kamcev (University of Michigan)

Ion exchange membranes (IEMs) that can discriminate between different counter-ions could enable separation applications such as lithium extraction, scalant removal, and nitrate recovery. Better understanding of ion partitioning and diffusion in IEMs contacted by mixed salts could guide the design of more selective IEMs. While there have been significant advancements in understanding single electrolyte transport in IEMs, studies probing counter-ion selectivity in IEMs with mixed salts are scarce. The water content and charge density of IEMs significantly influence counter-ion partitioning and diffusion in IEMs. However, the importance of these parameters in determining the counter-ion selectivity of IEMs has not been systematically probed. Additionally, most studies that have investigated mixed ion transport in IEMs utilize commercial membranes, which limits the development of structure/property relationships as the structures of these membranes are not controlled. In this study, a set of cross-linked cation-exchange membranes (CEMs) and anion-exchange membranes (AEMs) with varying water content and high charge density were synthesized via free radical polymerization. Equilibrium counter-ion sorption within the IEMs was measured with monovalent/monovalent as well as monovalent/divalent salt mixtures with total concentrations and equivalence fractions varied over a broad range. Notably, lowering the water content of the IEMs significantly increases both monovalent/monovalent and monovalent/divalent counterion selectivity in AEMs, whereas only slight differences are observed in CEMs. This behavior is likely due to specific ion effects, which are significantly strengthened in AEMs under confined environments of low water content membranes. In CEMs, counter-ion selectivity is mostly governed by electrostatics and differences in ion hydration. Results were interpreted within the framework of the recently developed Donnan-Manning model for mixed ion partitioning in IEMs.

4:40 PM

Oral 73 - Preparation of lithium-ion selective mixed matrix membrane for lithium recovery from real brine solution via diffusion dialysis process

Sandip Pal (University of California Los Angeles)*; David Jassby (UCLA); Eric Hoek (UCLA)

The global lithium market size was valued at USD 6.83 billion in 2021 and is expected to boom at a compound annual growth rate of 12.0% from 2022 to 2030. The increasing demand for lithium, a critical component in energy storage technologies, has intensified the search for efficient and sustainable methods of lithium recovery from brine sources. World lithium reserves are about 14 million tons, mostly 70–80% stored in Salt Lake brine, geothermal water, and solid lithium contained in lithium ore. The small world of lithium is expanding due to the emergence of electric vehicles. An energy efficient technology is required to recover lithium from the brine, and Diffusion Dialysis (DD) with ion exchange membranes (DD-IXM) has emerged as a promising technology for selectively extracting lithium ions from brine solutions. This study investigates the feasibility and optimization of lithium recovery using DD-IXM. Here, we report thermally crosslinked poly (2,6-dimethyl-1,4-phenylene oxide) (PPO) based lithium-selective anion exchange membranes (AEMs) by modification with Li-selective nanomaterials. Synthesized composite AEMs have been characterized for their physical, chemical, and morphological structures, as well as the uniform dispersion of lithium-selective nanomaterials throughout the membrane matrix. The fluxes of Li were found to be higher compared to Mg and Ca using the DD process by MM006 composite membrane. Interestingly, selectivity factors are exceptionally higher, e.g., ≥8, ≥250 & ≥150 for Li/Na, Li/Mg & Li/Ca, respectively. The results are pronounced for fabricated composite membranes for Li recovery via DD process. Thus, we can objectively estimate that the idea of the selective extraction of Li from a mixture of a solution containing other ions using a DD process was validated.

5:00 PM

Oral 74 - Mixed binary alkali halide salt transport in PEGDA

Aubrey Quigley (The University of Texas at Austin)*; Everett S Zofchak (The University of Texas at Austin); Kate Dycha (The University of Texas at Austin); Nathaniel Lynd (The University of Texas at Austin); Benny Freeman (The University of Texas at Austin)

Membrane research has recently seen a push towards aqueous solute recovery such as lithium salts from brines, rare earth elements from electronic wastes, and metals from mining effluents. However, a current challenge in the field is separating ions of like charge, size, and hydration. Understanding how ion transport is affected be the presence of other ions in solution can help advance the design of membranes tailored with specific ion affinities. Here, we use a model cross linked poly(ethylene glycol) diacrylate membrane to examine how the permeability coefficients of binary alkali halide salt mixtures vary as a function of external salt mole fraction. We see that, at constant ionic strength, mixed salt permeability is largely governed by mixed salt partitioning whereas mixed salt diffusion stays relatively constant due to minimal change in water volume fraction. Further, starting from the Nernst-Planck framework, we have derived a model for predicting permeability coefficients based off of a thermodynamic model for predicting partition coefficients, the Mackie-Meares model for predicting ion diffusion through polymers, and the Flory-Rehner model for predicting membrane water volume fraction. This new model, with no adjustable parameters, shows great agreement in conditions with minimal ion-ether oxygen coordination (i.e., Na+, Cl-, and Br-). This advancement in predicting permeability coefficients provides a promising avenue for enhancing membrane design tailored towards ion-ion selectivity.

5:20 PM

Oral 75 - Tuning ionic selectivity in nanoporous graphene membranes using atomic layer deposition and polyelectrolyte assembly

Simar K Mattewal (MIT)*; Rohit Karnik (MIT)

Ion-ion separations are crucial for resource recovery. Recovering rare earths and lithium from acidic leachates, are examples critical for clean energy infrastructure. Currently these ions are recovered through solvent extraction, which is solvent and energy intensive. Membranes, specifically nanoporous graphene have potential of concentrating ionic solutions and separating different ions. Being atomically thin they can achieve high permeances, however developing selective pores in these membranes is a challenge. In this work, graphene pores are tailored using atomic layer deposition (ALD) and polyelectrolyte assembly (PEA) to enhance ionic selectivity.

Nanoporous graphene membranes on polyimide support are tested with standardized solutions containing 10+ metal ions of different valences. The transport of ions through the membranes is studied in both diffusion and pressure-driven modes. The pores in these membranes are tailored by depositing hafnium oxide layers using ALD and decorated with positive and negative polyelectrolytes, introduced on either side of the membrane.

Transport of ions through nanoporous graphene followed size-dependent trends with the hydrated radii of the ions. ALD seals the defects on graphene and enables selective transport, leading to an alkali-rare earth selectivity of 7. ALD post-pore creation tightens the pore size distribution and increases the alkali-rare earth selectivity by 7 times. PEA across nanoporous graphene increases the alkali-rare earth selectivity to 125. Selectivity between ions of the same valence is also observed, for example a K+/Li+ selectivity of 10. The boost in selectivity in the graphene-polyelectrolyte system is investigated with different combinations of polyelectrolytes and lithium recovery is demonstrated on synthetic black mass solutions.

Modification of pores in nanoporous graphene through ALD and PEA, as shown in this work, opens the design space for tuning ionic transport through atomically thin membranes.

5:40 PM

Oral 76 - Lithium Selective Artificial Ion Channels for Lithium Separation

Harekrushna Behera (The University of Texas at Austin)*; Tyler J Duncan (The University of Texas at Austin); Laxmicharan Samineni (University of Texas at Austin); Hyeonji Oh (University of Texas at Austin); Venkat Ganesan (The University of Texas at Austin); Manish Kumar (University of Texas at Austin)

Lithium (Li) has been identified as a critical mineral, given its substantial demand in the manufacturing of high-energy-density rechargeable batteries, particularly for electric vehicles and grid energy storage. Current Li-separation methods encounter limitations in terms of throughput, energy consumption, and are known to adversely affect the environment. Separating lithium ions from other more abundant mono- and divalent metal ions can be challenging due to their similar size and/or charge. To address these challenges, we have proposed an affinity-based biomimetic separation technique that utilizes specific ligands designed for the selective separation of Li. Here, we present the design of a class of bioinspired supramolecular membrane channels that exhibit higher selectivity towards transporting lithium ions, compared to the commonly found monovalent ions (Li+/M+ ~100, M+= K+, Na+) and divalent ions (Li+/Mg2+ ~500, Li+/Ca2+ ~230) during electrochemically driven transport measurements. The ion transport/rejection mechanism was investigated further through molecular dynamics simulations. This mechanism can be attributed to the higher ion dehydration, and the associated energetic penalty, of monovalent ions at the ligands near the pore entrance, leading to their exclusion. Additionally, divalent ions are excluded due to size exclusion at the central pore. The unique ability of lithium to maintain its hydration shell and exhibit minimal interaction with the channel contributes to its faster transport. The enhanced selectivity of this channel towards Li ions compared to mono- and divalent ions makes it a promising candidate for constructing environmentally friendly and energetically efficient lithium separation membranes

Abstracts – Oral Presentations, Tuesday Morning

Presenting authors are indicated by an asterisk*.

Membranes for Carbon Capture I

8:30 AM

Oral 77 - Leveraging competitive sorption in microporous polymer membranes to enhance CO2- and H2S-specific gas separation performance

Pablo A Dean (Massachusetts Institute of Technology)*; Katherine Mizrahi Rodriguez (MIT); Yifan Wu (Massachusetts Institute of Technology); Zachary Smith (MIT)

Polymer membranes are promising for CO_2-based gas separations due to their processability, mechanical strength, and manufacturability. Recent advancements in polymers with permanent porosity (termed "microporous") have indicated that a phenomenon known as competitive sorption can be used to enhance separation performance. This presentation will focus on the fundamentals and practical implications of leveraging competitive sorption to enhance performance in application-relevant and multi-component gas mixtures.

Recent studies have reported that an amine-functional polymer of intrinsic microporosity (PIM-NH2) has CO_2/CH_4 selectivities for mixture experiments that are 260% higher than selectivities predicted from pure-gas calculations. This study seeks to investigate this unusual emergent phenomenon to evaluate if it is a generalizable feature of certain materials, and to evaluate how other strongly-sorbing species may influence performance. Results will be presented on the effects of amine functionalization on CO_2- and H_2S-selective gas transport in microporous polymer materials, including PIM-1 and a new family of microporous polymers within the poly-aryl ether (PAE) subclass. Sorption experiments demonstrate that acid gases exhibit higher affinity for the basic amine moiety relative to unfunctionalized chemistries, resulting in higher uptake despite having lower fractional free volume. The effects of competitive sorption are also investigated through mixed-gas permeation experiments, including ternary ($CO_2/CH_4/H_2S$) and binary (CO_2/CH_4 and CO_2/N_2) experiments to quantify competitive effects between CO_2 and H_2S relative to the less polarizable CH_4 and N_2 diluents. This work aims to provide an understanding of competition in microporous polymers that can be used in the design and evaluation of future membrane synthetic approaches, with a special emphasis on CO_2-focused mixtures that are relevant for many applications in the chemical industry.

8:50 AM

Oral 78 - Amine Functionalized Supported Ionic Liquid Membranes (SILMs) for CO2/N2 Separation

Antoine Chamoun-Farah (University of Texas at Austin)*; Austin Keller (University of Texas at Austin); Mariam Balogun (University of Texas at Austin); Louise Canada (University of Texas at Austin); Joan Brennecke (The University of Texas at Austin); Benny Freeman (The University of Texas at Austin)

Supported ionic liquid membranes (SILMs), containing aprotic heterocyclic anion ionic liquids (AHA ILs) in an inorganic inert support, exhibit excellent CO_2/N_2 perm-selectivity, reaching as high as 640 at 35° C and 0.03 bar CO_2, conditions similar to post-combustion carbon capture (PCCC) from a natural gas power plant. A Fickian model fit to the experimental data estimates CO_2 permeability at direct air capture (DAC) pressure conditions of 10,400 barrer and a CO_2-N_2 perm-selectivity of 4,000 for the best performing IL, triethyl(octyl)phosphonium 4-bromopyrazolide ([P2228][4BrPyra]). The effect of cation size on permeance is studied for the [2-CNPyr]- anion, and it was determined that the smaller [P2228]+ cation outperformed the [P66614]+ cation because its ILs had a higher carrier concentration and reacted with CO_2 more readily. The most important criterion for high selectivity is a large equilibrium constant for binding between the IL and CO_2, which results in high CO_2 solubility. ILs with smaller molar volumes and without any fluoroalkyl chains enhance N_2 rejection. Low viscosity and high IL molar density also enhance CO_2/N_2 perm selectivity and CO_2 permeance. In addition, preliminary humidified mixed-gas results, down to 400 ppm CO_2 (balanced N_2) will be presented for phosphonium-based ionic liquids with

AHAs.

9:10 AM

Oral 79 - Effect of Viscosity of Deep Eutectic Solvent on CO2 Capture Performance in an Energy Efficient Membrane Contactor Based Process

Syed Islam (Oak Ridge National Laboratory)*; Zachary Coin (Oak Ridge National Laboratory); Shailesh Singh Dangwal (Oak Ridge National Lab); Mary Irwin (University of Tennessee at Knoxville); Thomas Knight (Vanderbilt University); Ramesh Bhave (Ornl); Gernot Rother (Oak Ridge National Laboratory); Jacek Jakowski (Oak Ridge National Laboratory); Arif Arifuzzaman (Oak Ridge National Laboratory); Vera Bocharova (Oak Ridge National Laboratory); Robert Sacci (Oak Ridge National Laboratory); Ilia Ivanov (Oak Ridge National Laboratory); Jingsong Huang (Oak Ridge National Laboratory); David Sholl (Oak Ridge National Laboratory)

Greenhouse gas contributions to climate change have driven intense interest in separation of CO2 from wet flue gas streams. Deep eutectic solvents (DES) are an emerging class of highly selective CO2 absorbents. A prototypical DES, reline, is a mixture of choline chloride and urea. Reline is a thermally stable, non-toxic, and biodegradable solvent with negligible volatility and inexpensive. We demonstrate a scalable, and energy-efficient hollow fiber membrane contactor (HFMC)-based process using a green solvent for CO2 capture. This process uses reline in a HFMC to provide close interfacial interactions and contact between the DES and CO2. This approach overcomes disadvantages associated with direct absorption in DES and could potentially be applied to a variety of solvent-based CO2 capture methods. Commercial low-cost polymer hollow fiber membranes were evaluated for CO2 capture with reline. From a mixed gas containing N2 and CO2, the DES-based HFMC separated CO2 with a purity of 97 mol%. The effect of viscosity of reline on the CO2 capture performance was investigated by adding water to reline. By adding 30% water to reline, its viscosity reduced from 1600 centipoise to about 20 centipoise. The reduction of reline viscosity led to CO2 flux increased by a factor 5 in the membrane contactor system due to enhanced absorption and desorption kinetics. In-situ Fourier transform infrared (FTIR) measurements combined with density functional theory (DFT) based molecular dynamics simulations revealed that reline absorbs CO2 by physical absorption without forming a new chemical compound, and CO2 separation by reline occurs via pressure swing mechanism. This research provides fundamental insights about physical solvent-based separation processes and a pathway towards industrial deployment.

9:30 AM

Oral 80 - Incorporating phenyl groups in polyorganosilica membranes to improve hydrothermal Stability for H2/CO2 Separation

Vinh Bui (University at Buffalo)*; Varun Reddy Satti (University at Buffalo); Elizabeth Haddad (University at Buffalo); Ameya Tandel (State University of New York, Buffalo, NY); Narjes Esmaeili (The State University of New York at Buffalo); Sai Srikar Chundury (University at Buffalo); Lingxiang Zhu (National Energy Technology Laboratory); Haiqing Lin (The State University of New York at Buffalo)

Polyorganosilica (POSi) membranes can be rapidly prepared by scalable oxygen plasma treatment of polydimethylsiloxane (PDMS)-based thin-film composite (TFC) membranes, and they exhibit excellent H2/CO2 separation properties. However, they are subjected to hydrothermal instability, where simultaneous hydrolysis and condensation of the siloxane network decreases both H2 permeance and H2/CO2 selectivity. In this study, we incorporate hydrophobic phenyl organic groups into POSi membranes to improve membrane hydrothermal stability while retaining attractive H2/CO2 separation properties. Particularly, poly(methylphenylsiloxane-co-dimethylsiloxane-co-methylhydrogensilxoane) (poly(PMS-co-DMS-co-MHS)) membranes are exposed to 2-minute oxygen plasma treatment at room temperature to prepare an ultrathin (<10 nm) phenyl-decorated-POSi top layer, as confirmed by XPS and SEM. The effect of phenyl content on the H2, CO2, N2 and CH4 transport properties at 100oC are thoroughly examined and interpreted using resistance in series models assuming a thin silica layer on top of the polysiloxane membrane. For example, increasing the PMS content of 0 to 66 wt% increases H2 permeance from 150 to 190 GPU but decreases H2/CO2 selectivity from 78 to 38 at 100oC.

However, the addition of phenyl groups drastically enhances POSi membrane hydrothermal stability. For instance, after exposure to saturated water vapor at 100 oC for 12 h, H_2/CO_2 selectivity decreases by as much as 74% for the PDMS-derived POSi membrane but only 17% for the membrane derived from the copolymer containing 66 wt% PMS. The effect of the water vapor exposure on the chemical composition change of the membranes will be elucidated. The long-term stability of the phenyl-POSi membranes will be discussed to showcase their potential for practical H_2 purification and pre-combustion CO_2 capture.

9:50 AM

Oral 81 - Effect of Film Thickness and Temperature on the CO2 Sorption in Poly(ethyleneimine)

John Hoffman (NIST); Avery Baumann (NIST); Christopher M Stafford (NIST)*

Poly(ethyleneimine) (PEI) is an attractive material for CO_2-based carbon capture due to its high amine content (1 nitrogen for every 2 carbons). Hyperbranched PEI, specifically, is appealing due to its low glass transition temperature (Tg), which facilitates diffusion of CO_2 through the polymer. However, due to its viscous nature, hyperbranched PEI is typically imbibed into porous silicates to form a solid sorbent material. This coating process, while practical, leads to poorly defined PEI film thickness and local variations in PEI concentrations within the porous support. To understand fundamental aspects of CO_2 uptake and release within PEI-based sorbents, we propose to study planar PEI thin films, which allows us to explore the effect of film thickness and temperature on CO_2 sorption. In this presentation, I will discuss our measurements of adsorption/desorption at a function of temperature, humidity, and CO_2 concentration for different PEI film thicknesses using tandem quartz crystal microbalance (QCM) and polarization modulation-infrared reflection-absorption spectroscopy (PM-IRRAS). Uptake is shown to increase as temperature increases from 25 °C to 55 °C for samples > 25 nm, while a decrease is seen for films < 25 nm. CO_2 sorption at lower temperatures appears to be regulated by reaction kinetics, while CO_2 diffusion and mobility appear to be more important at higher temperatures. Coupling these two measurement techniques provides a unique testbed for quantification of mass uptake via QCM and spectral identification of reaction products via FTIR.

10:10 AM

Oral 82 - 100% Carbon Capture from a Carbon Dioxide-Argon Flue Gas using Polaris Membrane

Pingjiao Hao (Membrane Technology & Research Inc.)*; Tim Merkel (Membrane Technology & Research Inc.)

Natural gas-fired reciprocating engines are increasingly deployed for reliable power to balance renewables and are responsible for 70 million tonnes CO_2 emissions annually. The Argon Power Cycle (APC) is an innovative, ultra-efficient, flexible, and affordable power generation technology that uses a closed loop internal combustion reciprocating cycle working with argon as its working fluid. In this study, a Polaris membrane system was designed, built, and tested to capture 100% of the CO_2 generated by a 0.5 MWe APC system. The membrane system produces an argon-rich stream that is recycled to the combustion engine and a CO_2 concentrate stream that is ready to be used or sequestered.

One unique characteristic of the flue gas from APC system is a mixture of CO_2/argon, instead of a typical flue gas of mixture of CO_2/N_2. The APC system exhaust gas has 70-80% argon, 15 to 25% CO_2 and the pressure of the flue gas varies from 4 to 6 bar. This slightly elevated pressure, high CO_2 concentration of the APS flue gas and 30-50% single path capture rate make it well suited for membrane capture. In addition, the APS system rarely operates in a steady state, frequently ramping up and down to meet energy needs. This dynamic operation is challenging for most separation technologies, and again makes membranes an attractive option. Initially, lab tests on Polaris membranes with a mixture of CO_2/argon under the APC conditions were performed to determine membrane properties for designing the membrane system. A series of integrated system modellings were performed to optimize the control strategy for the dynamic operation of the APC carbon capture system. The three-month membrane-APC system field test was performed at the APC developer facility. The field test used commercial-sized Polaris modules and demonstrated that the system can capture 100% of the CO_2 from the APC flue gas and respond effectively when the APC system adjusts power output to meet variable electricity demand.

10:30 AM

Oral 83 - Thin Film Composite Membranes from Polymers of Intrinsic Microporosity for CO2 Separation

Sunshine Iguodala (Imperial College London)*

Polymers of intrinsic microporosity (PIMs) are a novel class of polymers with great promise for gas separation applications. The microporosity originates from the unique structure of the rigid polymer backbone of PIMs [1]. PIMs have so far largely been assessed as free-standing films; however thin film composite (TFC) membranes are crucial for achieving superior, commercially viable gas separation performance. The key to designing a high-performing TFC membrane is to develop a highly gas-permeable gutter layer and an ultrathin selective layer. The gutter layer serves to provide a flat, uniform surface to accommodate an ultrathin selective layer, overcoming the irregularities of the porous substrate that could otherwise hinder gas permeance. Defect-free and thin selective layers can boost permeance without sacrificing gas selectivity. Here, we first introduce the fabrication of a crosslinked aminopropyl-terminated polydimethylsiloxane (PDMS) gutter layer applied to a porous polyacrylonitrile (PAN) support. Our approach yields CO2 permeances exceeding 16,000 GPU, achieved through the careful control of thickness, thereby minimizing the partial filling of pores in the support layer during the fabrication process. We also extend the library of polymers that can be employed as selective layer materials, for example, the synthesis of fluorinated PIMs (F-PIMs) by superacid-catalyzed Friedel–Crafts reactions, with low water vapor sorption at high relative humidity for advanced gas separation processes. The content and distribution of fluorine can be adjusted easily, and an evaluation of the structure-property relationship will be provided. The TFC membranes achieve performances within the optimum region for industrial post-combustion CO2 capture as outlined by Merkel et al. [2] and are promising candidates for biogas separation.

Membrane Fouling

8:30 AM

Oral 84 - Optimizing electrospun composite-fiber patterns on nanofiltration membranes for improved biofouling resistance

Amos Taiswa (Montana Tech)*; Jessica Andriolo (Montana Tech); Jack Skinner (Montana Tech)

Biofouling on membrane surfaces remains a significant challenge in various separation processes, such as reverse osmosis, ultrafiltration, and microfiltration. The build-up of biofilm on water filtration membranes leads to decreased flux, higher energy consumption, and compromised long-term performance. Recent progress in bio-inspired designs, including biomimetic coatings and surface modifications inspired by physical anti-biofouling characteristics (such as super hydrophobicity and amphiphilicity etc.), shows potential in improving membrane resistance to biofouling. This study introduces a simple and biocompatible method for enhancing biofouling resistance on membrane surfaces using electrospun composite fibers. These fibers consist of a polymer blend comprising one hydrophobic component (polycaprolactone, PCL) and one hydrophilic component (polyethylene glycol, PEG). Under pressurized conditions with aligned fibers at ~10 µm spacings, a 49% reduction in biovolume is observed, whereas under no pressure conditions, biofilms tend to form. Patterns of PCL to PEG polymer blends, incorporating quorum sensing inhibitors (QSI), Urolithin A (200 µg/mL), demonstrate significant physical disruption and molecular interference in biofilm formation compared to pristine membranes. In a pressurized nanofiltration test, there is an 83.48% reduction, while electrospun fibers in a stagnant culture achieve an 84.15% reduction. The permeating flux also increased by 58.17 %. The authors also investigate the impact of fibers' sizes-spacings on biofouling, providing valuable insights into optimizing pattern design for enhanced nanofiltration performances.

8:50 AM

Oral 85 - Effect of Pattern Size Relative to Oil Droplet Size on Oil Fouling in Nanofiltration

Elina Yachnin (Technion - Israel Institute of Technology)*; David Jassby (University of California); Tamar Segal-Peretz (Technion- Israel Institute of Technology); Guy Z Ramon (Technion - Israel Institute of Technology)

Surface patterns are of interest to the membrane community after the success had in the biomedical industry with the Sharklet® pattern that mimics the form and function of shark skin. These surface patterns disrupt the fluid flow at the membrane surface, creating localized mixing that reduces fouling and make the membranes easier to clean. Fouling leads to a decrease in rejection and water transport through the membrane and will decrease the lifetime of the membrane. By adding surface patterns, we hypothesized that oil fouling will be decreased on the membrane surface. In this project, we investigated how the size of line-and-groove patterns affect fouling of ~10-micron sized oil droplets. The line-and-groove patterns are of 300 nm size, 10-micron size, and 50-micron size to look at patterns smaller than, equal to, and larger than the foulant size. Support membranes were patterned, and the nanofiltration (NF) polyamide layer was synthesized on top of the patterned supports. These patterned membranes are compared to commercially available NF membranes and non-patterned synthesized NF membranes. The membranes were challenged with three cycles of oil solution with cleaning and a water cycle after each oil solution. The flux recovery was calculated to determine the amount of reversible fouling. It was found that the pristine synthesized and nanopatterned sizes had the highest flux recovery of 91% and 89%, while the commercial and 10-micron had the lowest flux recovery of 78% and 83%. These results indicate that the pristine synthesized and nanopatterned membranes enhanced membrane cleanability.

9:10 AM

Oral 86 - The effect of alternating electric fields on membrane colloidal fouling

Elina Yachnin (Technion - Israel Institute of Technology)*; David Jassby (University of California); Tamar Segal-Peretz (Technion- Israel Institute of Technology); Guy Z Ramon (Technion - Israel Institute of Technology)

Colloidal fouling, where micro particulates are deposited on the membrane surface, often leads to reduced membrane performance and increased operational costs. Previous studies have shown that electric fields (EF) can reduce foulant accumulation near membrane-electrodes.1–3 However, the effect of alternating (AC) EF on colloidal fouling is not very well understood and received far less attention. We hypothesize that diffusivity mismatch between ions of asymmetric electrolytes leads to the formation of long-range electric fields under AC EF, which prevents particles from reaching the membrane. This phenomenon can be used as a membrane fouling mitigation technique.

In the present study, we investigated the effect of AC EF on colloidal deposition near an electrically conductive ultrafiltration membrane1, in-situ, using confocal microscopy. Carboxylated polystyrene microparticles and a diffusivity-asymmetric electrolyte (NaOH) were used as a model system, where hydroxide ions have higher diffusivity than sodium.

The results, presented in Fig. 1, show that the application of AC EF reduced colloidal deposition and, importantly, was affected by the AC frequency. We attribute this behavior to a negative long-range EF that formed due to negative charge excess of the highly diffusive hydroxide ions. These EFs repelled the negatively charged microparticles from the membrane surface. Interestingly, a frequency threshold for deposition mitigation was observed. Frequencies above the threshold may lead to higher hydroxide ion velocities and smaller displacements that eliminate the negative EF and increase colloidal deposition. This research provides insights into the ability and mechanism of AC EFs to mitigate colloidal fouling on ultrafiltration membranes.

9:30 AM

Oral 87 - Optimization of Fouling Resistant Polyelectrolyte Coacervate Membranes via Aqueous Phase Separation

Shao-Hsiang Hung (University of Massachusetts Amherst)*; Jessica Schiffman (University of Massachusetts Amherst)

Polymeric membranes fabricated from non-solvent induced phase separation (NIPS) are widely used in large-scale membrane-based separations, from the production of clean drinking water to kidney dialysis. However, the NIPS process relies heavily on toxic aprotic organic solvents, like N-methyl-pyrrolidine (NMP) and dimethylformamide (DMF). These solvents need to be removed before membrane application and require expensive recycling units due to strict environmental legislation and their concerns to human health. This study focuses on a new method of fabricating membranes created using a salinity-induced aqueous phase separation (APS) process that does not rely on the use of organic solvents. Polyelectrolytes are charged polymers that are soluble in aqueous solutions; however, by mixing them at the correct stoichiometry and in the presence of salt, counter-ions are released, and insoluble polyelectrolyte complexes (PECs) can be formed. By understanding the phase diagram of PECs, specific salt concentrations can be used to form a polymer solution that phase separates into polymer-lean (supernatant) and polymer-rich coacervate phases. In this study, we investigate how we can utilize the polymer-rich coacervate phase and their interactions with salt to form porous membranes. Experiments were started using the canonical system featuring the strong polyanion poly(sodium 4-styrenesulfonate) (PSS), and polycation poly(diallyldimethylammonium chloride) (PDADMAC) in the presence of potassium bromide (KBr) salt. The pore structure, separation performance, and fouling resistance of the membranes were evaluated using scanning electron microscopy (SEM), pure water permeance (PWP), dye removal tests, and microbial assays, respectively. Being able to manufacture chemically robust separation membranes with a range of pore sizes using the APS process without the use of any toxic solvents will be further investigated for a range of applications.

9:50 AM

Oral 88 - Mechanism of crystallization and membrane scaling mitigation via carbon nanotube-containing spacer in membrane process

Seongeom Jeong (Pusan National University)*; Sanghyun Jeong (Pusan National University)

Membrane technology facilitates phase separation and transport restriction through selective barriers. Many industries are interested in membrane technology for separating and rejecting small particles, such as reverse osmosis nanofiltration, and membrane distillation (MD). However, membrane technology faces several limitations, notably membrane scaling and fouling, which hinder its extensive industrial application. Membrane scaling and fouling are closely related to membrane performance deterioration. Crystallization often occurs due to membrane scaling, leading to the attachment of crystals on the membrane. To overcome these limitations, various flow promoters (e.g., spacers, modules, membrane patterns) are utilized. In a previous study, we demonstrated that 3D printing carbon nanotube (CNT)-containing spacer could enhance membrane flux and ion rejection in the MD process. We validated this phenomenon through computational fluid dynamics simulation. The CNT-containing spacer maintained a high flux (29 Lm-1h-1) until volume correction factor 4 was reached. Our focus then shifted to membrane scaling and crystallization when using the CNT-containing spacer in the MD process. We confirmed membrane flux and observed the membrane scaling phenomenon by the CNT-containing spacer in lab-scale MD processes. Additionally, we directly examined the effect of the CNT-containing spacer on crystallization. Through our research, we identified the mechanism by which the 3D-printed CNT-containing spacer mitigates the scaling problem by altering the crystallization phenomenon.

10:10 AM

Oral 89 - Atomic layer processes for membrane fabrication and surface modification

Brian Welch (Technion- Israel Institute of Technology); Ruoke Cai (Technion- Israel Institute of Technology); Assaf Simon (Technion); Zhenzhen Zhang (Helmholtz-Zentrum Hereon); Volker Abetz (Helmholtz-Zentrum Hereon); Tamar Segal-Peretz (Technion- Israel Institute of Technology)*; Clarissa Abetz (Helmholtz-Zentrum Hereon)

Atomic layer processes such as atomic layer deposition (ALD) and molecular layer deposition (MLD) are powerful techniques for growth of inorganic, hybrid, and organic materials conformally and with exceptional control over their nanoscale thickness. In recent years, these processes have emerged as new methodologies for controlling the membranes' interfaces and fabricating selective layers.
Here, I will discuss two examples for such processes. Al2O3 ALD was used to modify the interface of uni-porous self-assembled block copolymer membranes, tuning the pore size from ultrafiltration to nanofiltration. The Al2O3 interface also enables control over the pores' composition, leading to high selectivity of nano-pollutes. In addition, the inorganic-organic interface enabled unprecedent 3D characterization using TEM tomography and FIB-SEM slice-and-view, revealing hidden tortuosity and its connection to the membrane's performance.
MLD of vapor-phase m-phenylenediamine (MPD) and trimesoyl chloride (TMC) was used to grow layer-by-layer crosslinked, aromatic polyamides on top of polysulfone support membranes, producing highly smooth, thin-film composite (TFC) membranes. Key to these TFC membranes fabrication was the ability to grow MLD thin layers on top of the porous support despite the conformal nature of MLD. To overcome this challenge, we developed a sacrificial polyvinyl alcohol (PVA) pore filling process, that enabled thin polyamide layer growth at the top of the membranes. The PVA was later removed to yield TFC. The MLD polyamide layer showed similar or improved characteristics to interfacial polymerization polyamides, in terms of cross-linking degree, water permeability and salt rejection. Our research demonstrates a new avenue for polyamide TFC membrane fabrication, utilizing vapor-phase processes that enable exceptional control over the polyamide thickness and composition.

10:30 AM

Oral 90 - Fabrication of polymer brush grafted anti-biofouling ultra-filtration membrane using atom transfer radical polymerization

Kritika Jashrapuria (Indian Institute of Technology Bombay); Swatantra Pratap Singh (IIT BOMBAY)*

Water scarcity is one of the major global challenges, and membranes are increasingly employed for desalination and water purification as part of the solution. However, biofouling poses a substantial barrier to the effectiveness of all membrane-based technologies as it decreases the life of membranes, increases energy consumption, and decreases water flux. To date, significant efforts have been made to develop innovative techniques for modifying membranes to change their structure and surface characteristics for fouling mitigation. Recently, grafting zwitterion-based polymer brushes on surfaces as possible anti-biofouling materials have gained attention. Therefore, this study introduces a method for the development of a high-performance ultrafiltration (UF) membrane with the ability to withstand biofouling. This involved grafting fouling-resistant poly[2-(methacryloyloxy) ethyl dimethyl-(3-sulfopropyl) ammonium hydroxide] (pMEDSAH) zwitterionic polymer brushes onto a graphene oxide (GO) blended UF membrane. The surface-initiated atom transfer radical polymerization is used to graft the brushes, and this modification results in the development of membrane with increased hydrophilicity and enhanced anti-biofouling potential. The pMEDSAH grafted GO-PES membrane (PES-GO-PB) showed higher water flux and BSA rejection in dead-end filtration set-up. In addition, the pMEDSAH-modified membrane showed a ~90% reduction of biovolume in a dynamic biofouling condition. Thus, for water treatment, the pMEDSAH grafted GO-PES membrane may potentially offer a solution to biofouling.

Keywords: Anti-Biofouling, Hydrophilicity, Membranes, Zwitterionic Polymer brushes

Organic Molecules and Solute Selective Separations

8:30 AM

Oral 91 - Organic Solvent Reverse Osmosis Separations of Hydrocarbon-Alcohol Mixtures using Matrimid Membranes

Bennett D Marshall (ExxonMobil)*; Ryan Lively (Georgia Tech); Youngjoo Lee (Georgia Tech); JR Johnson (Exxon Mobil); David Moser (ExxonMobil)

The organic solvent reverse osmosis (OSRO) separation of alcohol / hydrocarbon mixtures with Matrimid membranes is studied via experiment as well as modelling with dry glass reference perturbation theory (DGRPT). It is shown that the separation of these mixtures is highly non-ideal, with strong maxima of the separation factors as a function of feed composition. The DGRPT model was fit to pure vapor sorption data, while the Fickian diffusivities were fit from pure liquid permeation data. From this pure component parameterization, the model successfully predicted the highly non-ideal separation behavior observed for mixtures of alcohols and hydrocarbons. Successful application of the model was crucial in the interpretation of the experimental data, conclusively showing that the non-ideal behavior observed in the separations was a result of competitive sorption effects.

8:50 AM

Oral 92 - Fluorine-Rich Poly(Arylene Amine) Membranes for the Separation of Sustainable Fuels and Hydrocarbons

Yi Ren (Georgia Institute of Technology)*; Ryan Lively (Georgia Tech); Sheng Guo (Nanjing University); Hui Ma (Nanjing University)

Low cost, renewable, and sustainable liquid fuels are critical components of the energy transition. Fischer-Tropsch (FT) syntheses can be used to convert carbon dioxide and water into a wide range of hydrocarbons such as paraffins, olefins, gasoline, and lubricants. However, the FT process creates a broad distribution of typically linear alkanes with some olefin byproducts. An electrically-driven, compact separation is needed to reduce the costs of renewable fuels and lubricants made from this process. We will discuss the design of a series of fluorine-rich poly(arylene) amine (FRPAA) polymeric membrane that can perform organic solvent reverse osmosis (OSRO) operations on many challenging hydrocarbon mixtures derived from the FT synthesis.

The FRPAA membranes were designed by incorporating fluorine-rich moieties in the structure of linear glassy polymers, which controlled solvent-induced polymer dilation to balance high separation efficiency with practical productivity. Utilizing this design, the FRPAA membranes not only exceed the current toluene/1,3,5-triisopropylbenzene trade-off line but also accomplish challenging FT derived separations such as the OSRO separation of hexane isomers, light olefin/paraffin, as well as the fractionation of light/heavy linear hydrocarbons. Technoeconomic analyses show that the FRPAA membranes, in a hybrid separation cooperating with distillation, can potentially reduce energy consumption and carbon emissions by 10-50% relative to stand-alone distillation.

9:10 AM

Oral 93 - Organic solvent nanofiltration and reverse osmosis performance of carboxylated PIM-1 in neat and inter-crosslinked mixed membrane configurations

Lucas Condes (University of Oklahoma)*; Matthew T Webb (University of Oklahoma); Will Box (University of Oklahoma); Michele Galizia (University of Oklahoma)

Organic solvent nanofiltration (OSN) and organic solvent reverse osmosis (OSRO) are promising candidates to reduce the energy burden of thermally-based separation processes, such as recovery of pharmaceuticals or fractionating hydrocarbons. In general, highly permeable materials have been reported, while highly selective performance remains difficult to achieve due to the aggressive environment. Additionally, the factors affecting their long-term stability remain essentially unknown.

Carboxylated PIM-1 (CPIM) is a more highly selective derivative of the widely studied PIM-1 precursor, but most studies on CPIM and its composites have revolved around gas separations. To fill this gap, in this work, we systematically investigate the OSN and OSRO potential of CPIM and composites based on CPIM and triptycene-isatin porous polymer networks (PPNs) to determine if their high selectivity and stability in gas separations translates to organic liquid separations. PPNs are crosslinked, organic 3-D networks that mix favorably with polymers to form mixed-matrix membranes (MMMs). However, their potential to improve selectivity, stability, or permeability in liquid separations remains unexplored.

We show that CPIM undergoes a radical-induced crosslinking reaction as low as 200 °C, occurring not only between CPIM chains, but also between CPIM and PPN, leading to the formation of inter-crosslinked, defect-free mixed matrix membranes (IcMMMs) exhibiting remarkable stability and high selectivity in gas and organic solvent separations.

For example, in an OSN test with polystyrene oligomers in toluene, neat CPIM has a toluene permeance of 0.042 L m-2 h-1 bar-1 with a MWCO of 200 g/mol, while thermally crosslinked CPIM has a toluene permeance of .046 L m-2 h-1 bar-1 with a MWCO as low as 162 g/mol.

The effects of crosslinking, PPN loading, solvent polarity, and solvent exposure time on the OSN and OSRO performance and its long-term stability are thoroughly investigated and discussed.

9:30 AM

Oral 94 - Volatile fatty acid selectivity during electrodialysis: role of solution composition on volatile fatty acid partitioning and mobility in an anion exchange membrane

Holly M Haflich (University of North Carolina- Chapel Hill)*; Joshua Singleton (University of North Carolina at Chapel Hill); Orlando Coronell (University of North Carolina at Chapel Hill)

Volatile fatty acids (VFAs) are short-chain carboxylic acids (2-5 carbon-chain) with versatile applications. Bio-based techniques can produce VFAs from organic waste, but VFAs must be selectively separated from other solutes for downstream application. VFAs are anionic at neutral pH, and therefore, electrodialysis (ED) can be used to separate VFAs from complex mixtures. Towards enhancing selective VFA recovery, VFA transport mechanisms in mixed-solute solutions during ED need to be better understood. Accordingly, we evaluated the effect of VFA identity on VFA permeability (P), partitioning (K), and mobility through an AEM during ED in the absence and presence of a competing anion. To do this, we measured the permeability and partition coefficients of VFAs (C2-C5) and chloride (Cl-) in single- and mixed-counterion solutions; mobility was calculated from P and K. We then compared the P, K, and mobility selectivities for single- and mixed-counterion conditions. Results showed that the permeability of shorter-chain VFAs was greater than that of longer-chain VFAs despite the higher partitioning of long-chain VFAs. For VFA/Cl- conditions, we observed an inverse trend in partition and mobility selectivities when comparing single- to mixed-counterion conditions. Specifically, the partitioning selectivities for VFAs over Cl- were lower in mixed-counterion solutions compared to single-counterion conditions, while mobility selectivities were higher in mixed-counterion solutions. Results suggest that the partitioning and mobility of any given VFA is decreased and increased, respectively, by the presence of a counterion with greater partitioning. Overall, VFA partitioning and mobility in the AEM, which govern VFA selectivity during ED, strongly depended on solution composition.

9:50 AM

Oral 95 - Membrane-based Ultrapurification of Isopropyl Alcohol

Kamalesh K Sirkar (New Jersey Institute of Technology)*; John Chau (New Jersey Institute of Technology)

Silicon wafers are subjected to an extraordinary amount of cleaning during semiconductor manufacturing to remove metallic impurities and organic compounds that are nonvolatile, semi-volatile or volatile. Ultrapure isopropyl alcohol (IPA) is needed since it is used to remove various organic contaminants. A vacuum membrane distillation process can be employed using two different types of membranes to produce ultrapure IPA. One type of membrane involves different polymeric nanoporous substrates having fluorinated surface coatings exposed to a mild vacuum. The second type of membrane is a subnanoporous perfluorocopolymer exposed to a substantial vacuum; we report here results using the second type of membrane. The condensed IPA quality was determined by gas chromatography-mass spectrometry (GC-MS) when the hot feed IPA at 40-45oC had a 2 ppm spike of C7-C30 hydrocarbons. The subnanoporous perfluorocopolymer membrane achieved in the IPA condensate 99.36% purification of the 143809 ppb of C7-C30 impurities injected into the IPA feed; further, there were no C19-C30 present in the permeate/condensate. The basis of the high purification achieved has been discussed and evaluated vis-a-vis other membrane processes such as reverse osmosis.

10:10 AM

Oral 96 - Co-Transport of Alcohols and Carboxylate Ions in Sulfonated Ion Exchange Membranes with PEGPEA as a Blocking Group

Pravin Parasakthi Aravindhan (Auburn University)*; Bryan S Beckingham (Auburn University); Jung Min Kim (University of Delaware)

Artificial photosynthesis devices such as photoelectrochemical CO2 reduction cells (PEC-CRC) could reduce CO2 to useable products such as methanol and carboxylates. However, product crossover through the membranes reduces device efficiency, and the transport of a particular product behaves differently in the presence of other products. Thus, it is desirable to understand membrane structure and the resultant transport-physiochemical property relationship to prevent this undesirable crossover without compromising ionic conductivity. Previously, our group observed increased acetate permeabilities on co-transport with fast diffusing methanol, and the inclusion of charge-neutral poly(ethylene glycol) phenyl ether acrylate (PEGPEA) could reduce this crossover. To further understand this behavior, a series of ion exchange membranes (IEMs) with poly(ethylene glycol) diacrylate (PEGDA) as a crosslinker, 3-sulfopropyl methacrylate potassium (SPMAK) as a sulfonated comonomer, PEGPEA as an uncharged comonomer were prepared. A total of 12 different compositions were prepared with 6 different SPMAK-to-PEGPEA ratios each totalling 32 mol% and 2 different crosslinker chain lengths (n = 10 and n = 13). Membrane permeabilities to alcohol (methanol) and carboxylates (formate, acetate, propionate) were carried out individually and in co-transport. Permeabilities vary as methanol > formate > acetate ≈ propionate. Permeabilities and ionic conductivities decrease with lower PEGDA chain length (n = 10). Total water, intermediate water, and non-freezable water increase with SPMAK content from 0 to 32 mol%. In the future, sorption desorption experiments with methanol and carboxylate will be carried out to understand varying structures toward solute transport and further our understanding of these structure-physiochemical-transport relationships.

10:30 AM

Oral 97 - Fundamental investigation about the effect of chemical structure of foulant on fouling behavior of polyamide membrane in organic solution

Atsushi Matsuoka (Kobe University)*; Ryouhei Aso (Kobe University); Eiji Kamio (Kobe University); Yasunao Okamoto (Kobe University); Tomohisa Yoshioka (Kobe University); Keizo Nakagawa (Kobe University); Hideto Matsuyama (Kobe University)

Organic solvent nanofiltration (OSN) is a separation process that has gained significant attention due to its energy efficiency. Many studies have successfully developed membranes that exhibit excellent solvent resistance, high permeance, and high model solute rejection. However, for the practical use of these membranes in industrial processes, membrane fouling can be a serious issue. Unfortunately, there are few reports about membrane fouling in organic solutions. Therefore, to develop membranes with high fouling resistance, it is important to investigate the factors that control membrane fouling in organic solutions.

To investigate the controlling factors of fouling, OSN tests were performed using a polyamide membrane and hexane solutions of several types of foulants with different functional groups. Differences in fouling behavior depending on the foulant structure were then investigated. The foulants used in the study were amine-, carboxy-, or hydroxy-functionalized straight saturated aliphatic compounds. The results showed that the decrease in permeation flux increased in the order of -NH2 > -COOH > -OH, indicating that the type of functional group affects fouling in organic solvents.

Additional studies using a quartz crystal microbalance revealed that the differences in the fouling properties for each functionalized aliphatic compound are due to the differences in the adsorption amount of each aliphatic compound on polyamide membranes. Furthermore, it was suggested that amino groups and carboxy groups interact strongly in hexane. Based on these results, it would be effective to reduce the number of amino groups and carboxyl groups on the membrane surface to produce polyamides with high fouling resistance.

Membrane Distillation and Pervaporation

8:30 AM

Oral 98 - Long-term Operation of Pilot-scale Membrane Bioreactor – Membrane Distillation for Potable Water Reuse

Luke Presson (University of Arizona)*; Michael Hegetschweiler (University of Arizona); Andrea Achilli (University of Arizona)

The implementation of membrane distillation (MD) is limited by the lack of knowledge on scaling and wetting resistance in long-term, continuous operation. In this study, a pilot-scale vacuum-assisted air gap MD (V-AGMD) system is operated in series with a membrane bioreactor (MBR) that uses municipal wastewater as the feed solution. The wastewater is treated to non-potable standards by the MBR subsystem and then treated to potable standards by the MD subsystem. The V-AGMD system operated at 75% recovery and produced 600-700 L/day of water for approximately three months. A decline in water flux due to fouling and scaling was observed but could be mitigated with routine chemical cleaning procedures. Cleaning with hydrochloric acid (HCl) was effective at recovering water flux, but after one month of weekly acid cleaning, some irreversible fouling and scaling remained. Further chemical cleaning procedures were tested including chlorination and the use of chelating agents. Chlorination was ineffective due to the lack of organic fouling while the chelating agent (Ethylenediaminetetraacetic acid, EDTA) recovered water flux by the removal of gypsum. Despite the evidence of membrane scaling, the water quality of the distillate remained high. The rejection temporarily decreased during the second month of operation but could be recovered by cleaning with EDTA. This study demonstrates that high water quality and water flux can be maintained in long-term MD operation with targeted chemical cleaning. Scalants like calcium carbonate can rapidly decrease the water flux with minimal effects on water quality, while gypsum scaling develops more slowly but instigates membrane wetting. Therefore, frequent acid cleaning and occasional cleaning with a chelating agent is recommended when utilizing MD for water reuse.

8:50 AM

Oral 99 - Combined chlorine dioxide–membrane distillation for the treatment of produced water

Mahdi Malmali (Texas Tech University)*; I-Min Hsieh (Texas Tech University)

The persistent challenges of fouling and scaling have hindered the effective implementation of MD for large-scale and industrial applications. In this presentation, I will discuss our systematic investigation, using chlorine dioxide combined with induced air flotation (ClO_2-IAF) as a pre-treatment technique of PW prior to treatment with MD. Initially, the optimization of ClO_2 generation, based on sodium chlorite and hydrochloric acid, was conducted in order to maximize on-demand ClO_2 production. The highest production yield of 18.4% was achieved with 4 wt.% $NaClO_2$ and 20 wt.% HCl solutions, with a molar ratio of 1:1.25. Subsequently, two actual PW samples were pretreated, and removal efficiencies for total suspended solids (TSS), turbidity, iron, and total organic carbon (TOC) were thoroughly investigated by varying the ClO_2 dosage from 6 to 91 mg/L. The ClO_2-IAF pre-treatment demonstrated TSS and turbidity removals exceeding 90%, with TOC removal nearing 55%. The PW constituents, such as benzene, toluene, ethylbenzene, xylene (BTEX), and total petroleum hydrocarbons (TPH), were analyzed and quantified throughout the cascade of the treatments. Our analysis suggests that the volatile hydrocarbons such as BTEX were primarily eliminated through air flotation, while they could easily pass through the MD membrane. The saturated hydrocarbons like TPH were not affected either by aeration or oxidation but were retained by hydrophobic PTFE membrane in MD. The increasing contact time between the ClO_2 and constituents in the PW affected by air flow rate significantly improved the long-term stability of MD, showing twenty-two days of stable performance without requiring in-place cleaning, emphasizing the critical role of optimizing oxidation pre-treatment to mitigate fouling in MD. The fouled membranes were characterized by SEM/EDX, and contact angle confirmed the effectiveness of the ClO2-IAF.

9:10 AM

Oral 100 - Membrane Distillation and Crystallization for Treating Hydraulic Fracturing Produced Water

Wickramasinghe, Ranil (University of Arkansas)*

Water is a very valuable natural resource. Maximizing reuse of wastewater is essential to meet increasing demands for fresh water. Hydraulic fracturing oil and gas flow back and produced water is frequently highly impaired. Currently it is often deep well injected into geologically isolated formations. Maximizing recovery of this water for beneficial uses is essential to promote a circular economy. However multiple unit operations are needed if these wastewaters are to be recovered for beneficial uses. Electrocoagulation is considered a promising pretreatment technology. Herein, we have investigated the use of aluminium electrodes for electrocoagulation as a pretreatment operation. The effects of electrode arrangement, applied current, reaction time, pH, and inter electrode distance on the quality of the treated water have been investigated.

The results obtained here indicate that electrocoagulation can obtain good removal of turbidity (95%), total suspended solids, TSS (90%), and total organic carbon, TOC (69%) by carefully choosing the reaction conditions. The performance of the electrocoagulation process depends strongly on the quality of the feed water. The viability of a practical continuous electrocoagulation process will depend on the volume (footprint) of the reactor which in turn will depend on operating conditions and the quality of the feed water.

9:30 AM

Oral 101 - Advances in the Characterization and Digital Reconstruction of 2-D and 3-D Pore Structure

Jay Werber (University of Toronto)*; Sima Zeinali (University of Toronto); Hooman Chamani (University of Toronto); Arash Rabbani (University of Leeds); Patrick Lee (University of Toronto); Jason R Hattrick-Simpers (University of Toronto)

Due to their high osmotic pressures, high-TDS brines remain out of reach for conventional reverse osmosis (RO), which has a maximum operating pressure of ~80 bar. Extensive research is on-going into enhanced RO processes such as high-pressure RO (HPRO), in which the maximum pressure would be extended to 120-200 bar, and variants of osmotically-assisted RO (OARO), in which a partially saline sweep solution lowers the osmotic pressure difference. For all enhanced-RO technologies, membranes will need to be developed with pore structures that are tailored to maximize compression-resistance while minimizing resistance to convective or diffusive flow. Detailed knowledge of membrane (support layer) pore structure is essential; however, current tools are highly flawed.

This talk will detail our on-going efforts in advancing the characterization of 2-D and 3-D pore structure. Our central goal is to combine conventional scanning electron microscopy (SEM) with digital tools to enable rapid generation of meaningful 3-D pore structures from easily accessible 2-D images. We are approaching this goal in two thrusts:

1) Improved methodology for 2-D structure generation. SEM images are 2-D, providing a gray-scale image of the top or cross-section of a membrane. For quantitative analysis, a binarized 2-D image is required, where the solid and void phases of the membrane are determined. We are exploring in detail the effect that experimental and digital methodologies have on the final binarized 2-D images.

2) Digital 3-D structure generation using data science. Here we are using these 2-D binarized images as

inputs to digitally generate 3-D structures that are statistically similar to the real structure. Our first efforts in this area were published in JMS (10.1016/j.memsci.2023.121673), with on-going efforts extending the approach to different membrane structures.

9:50 AM

Oral 102 - Fast and selective desalination using pressure-driven distillation with nanoscale air gaps

Anthony Straub (University of Colorado Boulder)*

Membranes that surpass existing limits in permeability and selectivity are urgently needed to enable energy-efficient water purification and industrial separations. Here, we discuss our team's development of desalination membranes that operate using pressure-driven vapor transport through nanoscale air gaps. We show that pressure-driven distillation membranes synthesized using an ultrathin hydrophobic coating on sub-100 nm pores have near-complete rejection (greater than 99.8%) of sodium chloride salts at operating pressures up to 34.5 bar. We also show the thickness of the air layer can be decreased down to 120 nm, and that such thin membranes can achieve water permeabilities per pore area of 8.9 kg m-2h-1bar-1 (five times higher than conventional RO membranes). The low aspect ratio capillaries allow us to probe the kinetics of liquid-vapor phase change, and we estimate the mass accommodation coefficient of water at conditions not previously accessible experimentally. We also, for the first time, validate predictions on the minimum aspect ratio of an entrapped air layer derived from thermodynamic wetting theory. Overall, this work demonstrates that nanoscale air layers can provide fundamental insights on phase change phenomena and offer a promising solution to challenges in sustainable water supply.

10:10 AM

Oral 103 - From surface wetting to internal wetting: wetting front progression in membrane distillation

Shounak G Joshi (University of Southern California)*; Amy Childress (University of Southern California); Allyson McGaughey (University of New Mexico)

Membrane distillation (MD) is a competitive process for treating challenging, high-salinity streams. While emerging pressure-driven processes such as high-pressure, osmotically assisted, and low-salt-rejection reverse osmosis are also being considered for high-salinity applications, these processes are complicated by membrane compaction, fouling, and heat management. MD can achieve high rejection and recovery at low pressures and with lower fouling rates as long as membrane wetting-resistance is maintained.

We recently used electrochemical impedance spectroscopy (EIS) to demonstrate that commonly used expanded polytetrafluoroethylene membranes typically operate in an internally wetted state without pore breakthrough. Here, we use EIS to show that at low feed salinities, movement of the wetting front is controlled by membrane hydrophobicity, membrane pore size, and transmembrane hydraulic pressure. However, at elevated feed salinities considerably below saturation, the depth of the wetting front is influenced by feed salinity.

Although wetting has traditionally been viewed as an on/off process (i.e., the membrane is either dry or wet), we show that it is complex. Our results show that wetting fronts move at different rates and pause/stop at different depths in the membrane, depending on operating pressure, feed solution chemistry, and membrane properties. We observe that in typical MD operation, wetting front progression is significantly slowed or terminated in close proximity to the distillate side of the membrane. This may be due to reduced scaling within membrane pores, fibril hydrophobicity at pore walls that provides internal wetting resistance, and/or observed changes in the microstructure of the membrane caused by elevated temperature & applied pressure. Our results have direct implications for understanding and quantifying membrane wettability in high salinity applications where maintaining a dry membrane is challenging but critical for process integrity.

10:30 AM
Oral 104 -

Abstracts – Oral Presentations, Tuesday Afternoon

Presenting authors are indicated by an asterisk*.

Desalination and Osmotic Processes II

1:00 PM

Oral 105 - Biomimetic membranes incorporating artificial water channels for high-performance water reverse osmosis desalination

Mihail Barboui (Institut Europeen des Membranes)*

Inspired by biological models, artificial water channels can be used to overcome the permeability/selectivity trade-off of traditional desalination membranes. We demonstrate that rational incorporation of I–quartet artificial water channels in composite polyamide membranes synthesized via interfacial polymerization, provide defect free biomimetic membranes with intrinsic water-to-salt permeability on the operational SWRO-BWRO/TWRO desalination pressure and high/medium/low salinity conditions. The best biomimetic composite membranes, which can be easily scaled for industrial standards (> m2), provide remarkable 99.5% rejection of NaCl and 91.2% boron rejection with water flux of 75 L m−2 h−1 at 65 bar applied pressure with a 35000 ppm NaCl feed solution at pH 8, representative of seawater desalination. This flux is more than 75% higher than that observed with current state-of-the-art membranes with equivalent solute rejection, translating into an equivalent reduction of the membrane effective area for the same water output and a ~12% reduction of the required energy for desalination.

1. M. Di Vincenzo, A. Tiraferri, V.-E. Musteata, S.Chisca, R.Sougrat, L.-B. Huang, S. P. Nunes, M. Barboiu, Nature Nanotechnol. 2021, 16, 190-196

2. L.-B. Huang, M. Di Vincenzo, M. Göktuğ Ahunbay, A. van der Lee, D. Cot, S. Cerneaux, G. Maurin*, M. Barboiu, J. Am. Chem. Soc. 2021, 143, 14386-14393.

3. M. Di Vincenzo, A. Tiraferri, V.-E. Musteata, S. Chisca, M. Deleanu, F. Ricceri, D. Cot, S. P. Nunes, M. Barboiu, Proc. Natl. Acad. Sci., 2021, 118(37), e2022200118.

1:20 PM

Oral 106 - Upcycled PVC Support Layer from Waste PVC Pipe for Thin Film Composite Nanofiltration Membranes

Atta Ur Razzaq (The University of Alabama)*; Milad Rabbani Esfahani (The University of Alabama)

Thin-film-composite (TFC) nanofiltration membranes represent the pinnacle of membrane technology in water treatment and desalination. These membranes typically consist of a polyamide (PA) selective layer and a membrane support layer, often constructed from commercially available materials like polyethersulfone (PES). However, there exists an alternative approach that involves the use of different polymers, preferably upcycled waste polymers, as a viable support layer for TFC membranes. Successfully implementing the upcycling of waste plastics into high-value support membranes can make a significant contribution to addressing the issue of plastic waste pollution. One of the main challenges associated with utilizing upcycled polymers as support layers is the potential impact on the polyamide selective layer of TFC membranes, subsequently affecting their performance in terms of permeability, rejection, and antifouling properties. In this study, we demonstrate the successful fabrication of TFC membranes with a support layer crafted from upcycled waste PVC pipe. We conducted a comprehensive investigation into the effects of upcycled PVC on the structural and physicochemical properties of the polyamide layer. The ultrafiltration (UF) support membranes were fabricated from waste PVC pipe via non-solvent induced phase separation (NIPS). Subsequently, a polyamide layer was synthesized atop the Upcycled PVC membrane using interfacial polymerization (IP). The physicochemical properties and performance of the TFC membranes with upcycled PVC support layer were compared with membranes with research-grade (RG) PVC and commercial PES support layers. The results unveiled that the TFC membrane with upcycled PVC support layer exhibited higher water permeability (18.2 LMH/bar), in contrast to RG TFC (15.5 LMH/bar) and PES TFC membranes (13.7 LMH/bar). Furthermore, the salt rejection capabilities of the upcycled PVC TFC membrane were competitive and well within an acceptable range.

1:40 PM

Oral 107 - Maximizing the Wetting Resistance of Fluorine-free Omniphobic Membranes for Hypersaline Wastewater Desalination

Yinchuan Yang (University of British Columbia)*; Sifat Kalam (University of British Columbia); Sadaf Shabanian (University of British Columbia); Kevin Golovin (University of Toronto); Yalei Zhang (Tongji University); Jongho Lee (University of British Columbia)

Membrane distillation (MD) equipped with omniphobic (non-wetting) membranes has found a niche in water reclamation from hypersaline industrial wastewater. The expanding ban on fluorocarbon-containing materials necessitates manufacturing omniphobic membranes from fluorine-free chemicals. Here, we examine the efficacy of representative non-fluorinated materials as surface coating agents for omniphobic membrane fabrication, and identify necessary mechanisms to attain a maximized wetting resistance using fluorine-free materials.

We first prepared MD membranes with different surface chemistries using a series of linear alkylsilanes and polydimethylsiloxane (PDMS) as representative fluorine-free, low surface energy materials. Membranes modified with a longer chain alkylsilane exhibited a lower surface energy and demonstrated a greater wetting resistance in direct contact MD experiments using feedwaters of various surface tensions. Despite the nearly identical surface energy measured for the longest alkylsilane and PDMS, PDMS-modified membrane exhibited an extended antiwetting performance as compared to the membrane treated with the longest alkylsilane. To elucidate the source of the distinctive wetting resistances, we examined the surface nucleation and condensation kinetics of the surfaces with the different surface chemistries via environmental scanning electron microscopy. Our analysis suggests that the membranes treated with long chain alkylsilanes contain surface defects (i.e., hydrophilic regions) whereas the high mobility of the PDMS effectively minimizes the defect exposure, slowing down the condensation and subsequent surface wetting. This work highlights that surface defect minimization is a crucial strategy, beyond surface energy reduction, for fluorine-free omniphobic membrane fabrication.

2:00 PM

Oral 108 - Direct Membrane Osmometry for Rapid Osmotic Pressure Measurements at High Molality

Mi Zhang (University of Connecticut)*; Jeffrey R McCutcheon (University of Connecticut); George Bollas (University of Connecticut); Edward Wazer (University of Connecticut); Wajeha Tauqir (University of Connecticut)

The accuracy of osmotic pressure measurements are important for predicting the performance of osmotic membrane systems such as RO, HPRO, FO, pressure retarded osmosis (PRO), and osmotically assisted reverse osmosis (OARO). State of the art osmometry involves a range of approaches, each tailored to specific applications. Vapor pressure osmometry (VPO) and Freezing point osmometry (FPO) are two of the modern commercialized osmometers that utilize the dewing point depression and freezing point depression in low concentrated solutions. Despite its fast measurement and repeatability, FPO is limited to for dilute biological and aqueous solution with molality <500 mmol/kg and inadequate for any suspended particulates that can act as nuclei for crystallization. While VPO cannot be used for volatile solutes like alcohols or organic solvents nor high molality solutions. Membrane osmometers on the other hand, provides a direct osmotic pressure measurement by allowing a net water flow from pure solvent through the membrane, into the compartment containing concentrated solutions during natural osmosis. Direct measurement techniques are more valuable to predicting osmotic membrane system performance as they measure the real "pressure" being exerted by the solution and avoid indirect thermodynamic assumptions that both VPO and FPO utilize. Current membrane osmometers are unfortunately unable to measure high pressures or take hours or days to process a single data point. In our work, we designed a novel customized membrane osmometer which not only possesses exceptionally truncated measurement time to within 5 minutes, but also demonstrates the osmotic pressure measurement ranges up to 2800 psi, while maintaining accuracy higher than 95% even for small monovalent salt such as sodium chloride (molality up to 4.5 M). To our knowledge, this work represents the most efficient and fastest measurements ever reported and could yield better prediction of membrane process performance.

2:20 PM

Oral 109 - XPRIZE Water Scarcity

Hossein Atoufi (XPRIZE)*

Water scarcity is an escalating global crisis that affects nearly one-fourth of the population, with predictions indicating that by 2030, demand will surpass supply by 40%. This looming crisis is predominantly impacting densely populated coastal cities and is exacerbated by the limited 0.5% of Earth's freshwater that is usable, strained further by rapid urbanization, population growth, and climate change. XPRIZE Water Scarcity, a $119 million, 5-year competition, aims to address these challenges by fostering innovative advancements in seawater desalination technology. As 70% of the planet is covered by oceans, which contain over 96% of global water resources, desalination offers a potential solution for transforming abundant seawater into a sustainable freshwater supply. However, existing desalination methods, while mature, have not seen significant innovation since the early 2000s and are often prohibitively expensive and environmentally unsustainable for widespread adoption. The competition is structured into three tracks to tackle these challenges: the development of a new desalination system capable of sustainably producing one million liters of potable water per day, the creation of novel membrane materials to enhance the efficiency and lifespan of reverse osmosis processes, and an ideas competition to reshape the global perception of water's value. Through these efforts, XPRIZE Water Scarcity seeks to catalyze a paradigm shift in desalination technology by making it more robust, affordable, and environmentally sustainable. This initiative not only aims to redefine the landscape of desalination on a global scale but also to ensure broader socio-economic equity and enhance global water management practices, creating a future where clean water is equitably and sustainably abundant.

2:40 PM

Oral 110 - Delamination Mitigation of Modified Bilayer Membranes using Single Slot Die Roll-to-Roll Manufacturing Process

Minwoo Jung (Georgia Institute of Technology)*; Rachel Kaplan (University of Kentucky); Isabel Escobar (University of Kentucky); Tequila Harris (Georgia Institute of Technology)

Polymer membrane fabrication using non-solvent induced phase separation (NIPS) method has been widely used. There are a wide variety of polymers that can be used to fabricate membranes using NIPS, such as Polyvinylidene fluoride (PVDF), Polysulfone (PSf), among others for water purification purposes. Furthermore, Polyamide (PA) and Polyimide (PI) are used for desalination process. Yet, current membrane products are mostly single layer with single function unless additive materials are added. With more layers, more functionalities can be added to the membrane. For instance, adding PA or PI layer to the PVDF or PSF membrane can add desalination function and water purification function. To utilize the functionality of each type of polymer for membrane, there are recent studies to create a bilayer structure of membrane that uses two different polymers for each layer.

A major limitation upon fabricating these materials, using one or more steps, is delamination between the top and bottom layers. To address these issues, in this study, we have developed a modified polymer system to mitigate the delamination of the bilayer membrane using a well-controlled coating process on a roll-to-roll system to fabricate the membranes. The bilayer membranes were fabricated using sacrificial material PVDF, PSf, and additive polymer. It has been found that delamination can be mitigated through enhanced material formulations that enhance layer adhesion, Figure 1. The delamination mitigation technique introduced in this work can be utilized to fabricate desalination membranes by adding a reverse osmosis membrane layer.

NAMS Awards

1:00 PM

Oral 111 - How click chemistry can enable design and investigation of facilitated CO_2 separation membranes

Joshua D Moon (University of Florida)*

Facilitated transport membranes (FTMs) use reversible interactions between CO_2 and ligands such as amines to increase CO_2 permeability and selectivity. There is a need to understand more fully the relationships between polymer functionality, architecture, and CO_2 sorption and diffusion in FTMs that use new amine chemistries, particularly in the presence of variable humidity. Unfortunately, identifying key structure-property relationships is hindered by synthetic difficulties of efficiently incorporating diverse chemical moieties into well-controlled polymer networks.

Tackling these challenges requires a more chemically modular and versatile synthetic strategy for preparing polymer membranes for CO_2 separations. The ability to control both spatial distribution and type of functional groups interacting with CO_2 through simple but quantitative post-modification methods could potentially circumvent challenges associated with the CO_2 upper bound tradeoff while also uncovering fundamental mechanisms of CO_2 interactions with grafted ligands in the presence and absence of water vapor.

We addressed these challenges by building on a synthetic strategy that uses active ester "click" chemistry to post-functionalize poly(ethylene glycol) diacrylate (PEGDA) copolymer membranes with different Lewis bases. We reveal how grafting density and pKa of different heterocyclic amine ligands (e.g., pyridine, imidazole, piperidine) affect gas solubility and diffusivity and corresponding activation energies in PEGDA-based FTMs, with CO_2/N_2 selectivities observed as high as 80 at 10 atm and 35 °C. We further evaluated CO_2/N_2 transport in both fixed and mobile carrier membranes by grafting ligands to networks and comparing with networks with unbound ligands. CO_2 and N_2 permeabilities are coupled with fractional free volume with ligand pKa playing an independent role. Ongoing work is investigating gas transport in membranes functionalized with multiple ligands and under controlled humidity.

1:20 PM

Oral 112 - Rational design, synthesis, and characterization of facilitated transport membranes exhibiting enhanced permeability, selectivity, and stability

Matthew T Webb (University of Oklahoma)*; Lucas Condes (University of Oklahoma); Harold Ly (University of Oklahoma); Michele Galizia (University of Oklahoma); Sepideh Razavi (University of Oklahoma)

Facilitated transport membranes (FTM) show superior performance over conventional polymer membranes over a short time frame, as they are susceptible to rapid performance decline due to photo/chemical aging. Additionally, the creation of mixed membranes (MMMs) by integrating silver nanoparticles and silver salts with polymer materials often leads to structural imperfections at the metal-polymer interface, diminishing selectivity. To address these challenges, a series of poly(amic- acid)s (PAAs) exhibiting systematically varied molecular weight and ether groups concentration were synthesized and used to fabricate, via the chelating-mechanism, silver nanoparticles in attempts to simultaneously i) achieve defect-free mixed matrix membranes (MMMs), ii) target CO_2/CH_4 and ethylene/ethane selective transport, and iii) reduce photochemical induced degradation. Individual and un-aggregated PAA-chelated silver nanoparticles were obtained and incorporated into a commercial polymer, Pebax 1657, to fabricate defect-free mixed matrix membranes.

The effect of the particle weight loading (0-4 wt%), and the PAA length and ether functional group concentration on the structure and performance of the resulting MMMs was systematically investigated. Remarkably, inclusion of only 2 wt% nanoparticles in Pebax 1657 enhanced CO_2 permeability by about 50% and CO_2/CH_4 selectivity by about 100% relative to neat polymer. A detailed analysis of the transport mechanism and energetics was performed to elucidate the molecular origin of the observed membrane performance. Finally, the newly synthesized materials were exposed to high pressure hydrogen over the course of 1,000 hours, to assess the long-term chemical stability of these materials.

1:40 PM

Oral 113 - Surface coatings for fouling-resistant membrane distillation

Kecheng Guan (Kobe University)*; Hideto Matsuyama (Kobe University)

Membrane distillation (MD) has prominent advantages such as treating high-salinity wastewater with a low-grade thermal energy, high salt rejection, and zero discharge. However, organic fouling and mineral scaling are two prominent challenges for hydrophobic MD membranes when used for practical applications. Commonly, improving organic fouling- and mineral scaling-resistance require oppositely enhanced wetting properties of membrane, thus is difficult to simultaneously realize dual resistance for one membrane. Here, we proposed to use underwater thermodynamically stable high-surface-energy hydrogel coating to modify the hydrophobic membrane with Janus structures comprising different surface energy. The underlayered structure meets the hydrophobicity requirements of the MD membrane, while the coating layer realizes dual resistance to organic and inorganic foulants.

A sodium alginate (SA)-based coating layer was deposited on a superhydrophobic polytetrafluoroethylene (PTFE) membrane to obtain a membrane with Janus surface energy via using the layer-by-layer (LbL) assembly through electrostatic attractive forces. The coating surface tended to be continuous when more coating cycles were performed. The successful coating of the PTFE membrane was confirmed by surface chemistry analysis. Fouling evaluations, including hexadecane fouling and gypsum scaling in the MD process, were then experimentally performed to confirm the potential of the PEI/SA material. The coating enabled the PTFE membrane with promising performance and longevity for the desalination and concentration of real seawater. The thin and continuous structure of the membrane coating could retain the anti-wetting property of the original membrane, and the high surface energy realized a robust hydration layer that prevented organic and inorganic foulant adhesion.

2:00 PM

Oral 114 - Ultrahigh Charge Density Ion-Exchange Membranes

David Kitto (University of Michigan)*; José Carlos Díaz (University of Michigan); Carolina Espinoza (University of Michigan); Jacquelyn Zamora (University of Michigan); Jovan Kamcev (University of Michigan)

Ion exchange membranes (IEMs) are dense polymer films functionalized with ionized moieties. IEMs are used extensively in electrochemical water and energy applications such as electrodialysis, fuel cells, flow batteries, and electrolysis. These technologies require exclusive and expedient transport of ions through IEMs, quantified as selectivity and conductivity. Unfortunately, a trade-off between these properties impedes membrane performance, where highly selective membranes have low conductivity and vice versa. Therefore, technologies using IEMs tend to be limited by one of these membrane properties. Using thermodynamic models, we have found that the polymer charge density defines the upper bound of the performance trade-off. So, to improve IEMs, we designed materials with higher charge densities than current membranes.

It is challenging to prepare high charge density membranes because charged polymers swell extensively with water. To retain control over both the water swelling and charge density, we synthesized various cross-linkable ionic monomers. By testing numerous cross-linking densities, cross-linker sizes, and membrane morphologies, we have systematically produced a library of ultrahigh charge density IEMs. To add context to these results, we also characterized the performance of 40 commercially available IEMs and surveyed properties from ~1,000 reported IEMs.

IEMs prepared from our monomers achieved ultrahigh charge contents (7-9 M), exceeding every charged membrane surveyed in the literature (<6 M). Following our model predictions, these materials have allowed us to completely re-define the upper limit of highly selective, highly conductive ion transport. With such favorable performance, ultrahigh charge density membranes promise significantly improved energy consumption for IEM-based technologies.

2:20 PM

Oral 115 - Tuning the selectivity of fouling resistant zwitterionic copolymer membranes

Sam J Lounder (ZwitterCo)*; Ayse Asatekin (Tufts)

Increased strain on our water economy necessitates responsible reuse, treatment, and discharge of various water/wastewater sources. However, current separation technologies are unfit to address a broad range of these separations, particularly those involving high concentrations of organic macromolecules, fats/oils, and other foulants. Self-assembling zwitterionic amphiphilic copolymers (ZACs) provide a robust and revolutionary platform for developing membranes with exceptional fouling resistance, enabling the first-time treatment of historically unfilterable industrial wastewater sources in a scalable and now commercialized technology.

This work describes novel methods to achieve fouling resistant ZAC membranes with customizable separation performance. Emphasis is placed on tuning self-assembled pore size through UV-based polymer cross-linking, a robust platform that enables the first-time use of ZAC membranes in applications such as water softening, selective ion removal, brackish water desalination, and ultrafiltration. We show that a wide range of pore sizes can be accessed through levers such as copolymer composition and UV-curing conditions. We explore the interaction-based ion selectivity of membranes possessing sub-nanometer pores and discuss implications on macromolecular design. We also demonstrate ultra-fast cross-linking achieved through thiol-ene click chemistry, providing a potential route for technology scaleup. Finally, we discuss new opportunities for translating novel chemistries to full-scale 80-40 modules through ZwitterCo's recently opened Technology Innovation Center (TIC).

2:40 PM

Oral 116 - Upcycling wastewater nitrate into ammonia fertilizer via electrified membrane

Jianan Gao (New Jersey Institute of Technology)*

The electrochemical upcycling of nitrate (NO3−) to ammonia (NH3) holds promise for synergizing both wastewater treatment and NH3 synthesis. Efficient stripping of gaseous byproducts (NH3, H2, and N2) from electrocatalysts is crucial for continuous and stable electrochemical reactions. This study evaluated a layered electrocatalyst structure using copper (Cu) dendrite to enable a high curvature and hydrophobicity and achieve a stratified liquid contact at the gas-liquid interface of the electrocatalyst layer. As such, gaseous product desorption or displacement from electrocatalysts was enhanced due to the separation of a wetted reaction zone and a non-wetted zone for gas-transfer. Consequently, this electrocatalyst structure yielded a 1.9-fold boost in per active site activity compared to that with a low curvature and high hydrophilic counterpart. Moreover, a high NH3 Faradaic efficiency of 90.9 ± 2.3% was achieved with nearly 100% NO3− conversion. This high-curvature hydrophobic Cu dendrite was further integrated with a gas-extraction membrane to construct an electrified membrane module, which demonstrated outstanding NH3 yield and separation from the real reverse osmosis retentate brine.

Advanced Understanding and Synthetic Efforts in Solute Selective Separations

1:00 PM

Oral 117 - Microwave dielectric relaxation spectroscopy: A technique to inform ion transport in hydrated polymer membranes

Geoffrey M Geise (University of Virginia)*

Providing sustainable supplies of purified water and energy is a critical global challenge for the future, and polymer membranes will play a key role in addressing these clear and pressing global needs for water and energy. Polymer membrane-based processes dominate the desalination market because they are more energy efficient than thermal desalination processes, and polymer membranes are crucial components in several rapidly developing power generation and energy storage applications that rely on membranes to control, selectively, rates of ion transport. Much remains unknown about the influence of polymer structure on basic intrinsic transport properties, and these relationships are important for designing next generation polymer membrane materials.

Key to controlling ion transport in membranes for water purification and energy applications is engineering interactions between ions and the solvated polymer matrix. Microwave dielectric relaxation spectroscopy is a tool that enables study of hydrated polymer membranes because water molecule dipoles can be probed in the microwave frequency region. This technique can provide unique insight into hydrated polymer dielectric permittivity (or dielectric constant) properties, which are critical in emerging thermodynamic modeling of ion transport in hydrated polymers. This presentation discusses the use of microwave dielectric relaxation spectroscopy to uncover structure property relationships in hydrated polymers and, more broadly, presents an overview of research aimed at further understanding fundamental structure/property relationships that govern small molecule transport in polymeric materials considered for desalination and electric potential field-driven membrane applications that can help address global needs for clean water and energy.

1:20 PM

Oral 118 - Insights into methods of determining selectivity of porous ultrafiltration membranes

Suzana Ivandic (UT Austin)*; Mostafa Nassr (UT Austin); Alishba Amjad (UT Austin); Ben Pedretti (MIT); Nathaniel Lynd (The University of Texas at Austin); Lynn Katz (The University of Texas at Austin); Benny Freeman (The University of Texas at Austin)

The selectivity of porous membranes is typically characterized using solute rejection tests, where membranes are challenged with dilute aqueous solutions of neutral solutes at operating conditions that minimize concentration polarization and fouling effects. The solute rejection profile and molecular weight cut-off (MWCO) are useful metrics in understanding pore size properties of a membrane. To obtain rejection profiles, solute concentrations are determined using techniques, including total organic carbon (TOC) and high-performance liquid chromatography (HPLC). Traditionally, single solute tests are thought to predict membrane characteristics most accurately, where a membrane is challenged with one molecular weight (MW) solute at a time from low to high MW. Since single solute methods are time-intensive, mixed solute tests have been utilized, where a mixture of several MW solutes challenges a membrane at once. However, there are no universally accepted operating conditions or standard methods used by membrane manufacturers or researchers, which leads to difficulty in pore size and pore characteristic comparisons. In this study, single and mixed solute tests as well as TOC and HPLC techniques are compared to help develop an easy, efficient, and validated method for porous membrane characterization. High MWCO polysulfone membranes are challenged with single and mixed polyethylene glycol solutes, where discrepancies in rejection curves obtained by TOC and HPLC are attributed to polymer dispersity. Utilizing chromatography, mixed solute rejection curves underpredicted MWCO compared to single solute curves, and mixed-solute concentration polarization and differential solute lag hypotheses are discussed as potential explanations. Understanding the differences between characterization techniques is key in developing a repeatable and universally accepted method for predicting membrane characteristics and further developing porous membranes.

1:40 PM

Oral 119 - Mixing and Matching: Amhiphilic Polyampholytes and Polyelectrolyte Complexes that Mimic Zwitterionic Behavior for Membrane Applications

Luca Mazzaferro (Tufts University); Ayse Asatekin (Tufts)*

Zwitterionic polymers, comprising anionic and cationic groups on the same repeat unit, can be extremely resistant to fouling, one of the largest obstacles to the broader adoption of membrane separations in many applications. Recent work has shown that Zwitterionic Amphiphilic Copolymers (ZACs) can create extremely fouling-resistant membrane selective layers whose properties and selectivity can be tuned by controlling their chemistry. However, the narrow range of commercially available zwitterionic monomers and difficulties in dissolving and processing resultant polymers limit their utility. This presentation will focus on how amphiphilic copolymers that incorporate a combination of anionic and cationic repeat units can behave similarly to ZACs and even surpass their performance. One such approach involves the use of amphiphilic polyampholytes (APAs), which contain hydrophobic, anionic, and cationic repeat units. Alternatively mixtures or consecutive layers of two amphiphilic copolymers, one cationic and one anionic, can create Amphiphilic PolyElectrolyte Complexes (APECs) that create nanochannels through pairing of charged repeat units. Both of these approaches result in membranes with excellent fouling resistance, coupled with tunable selectivity in roughly the nanofiltration range, prepared using scalable approaches. This "mixing and matching" approach can significantly broaden the range of chemical functional groups that can be incorporated into these membranes, opening the door for better control over selectivity and interactions for future applications.

2:00 PM

Oral 120 - Highly rigid and microporous polyimine membranes by interfacial polymerization for efficient hydrocarbon separations

Hiroki Fukuda (University of British Columbia); Jongho Lee (University of British Columbia)*

Hydrocarbon separations (e.g., crude oil fractionation) by organic solvent reverse osmosis (OSRO) is an emerging application of the membranes within a relatively low molecular weight cutoff (MWCO) range (<400 g mol−1). In this work, we have developed high-performance OSRO membranes through a scalable interfacial polymerization technique by integrating highly rigid polyimine chemistry and micropore-generating units. Compared with the conventional amide linkage which has been used for interfacial polymerization, the fully conjugated imine linkages are so rigid that the resultant polyimine membranes not only possess high microporosity but can withstand the swelling by organic solvents. These features significantly improved both toluene (kinetic diameter, dk = 5.9 Å) permeance and toluene/1,3,5-triisopropylbenzene (TIPB, dk = 9.0 Å) selectivity of polyimine membranes compared to the polyamide counterparts. In addition, introducing a shape-persistent triptycene unit could maximize the ultramicroporosity (< 7 Å) and the swelling resistance by showing an unprecedentedly high toluene/TIPB selectivity up to ~30. Ultimately, the developed triptycene-based polyimine membranes demonstrated excellent potential for hydrocarbon separations by a week-long 12-component mixture test under an industrially relevant cross-flow condition.

2:20 PM

Oral 121 - Real-Time, Non-Destructive Monitoring of Supported Liquid Membranes for Biochemical Recovery Using Electrochemical Impedance Spectroscopy

Hiroki Fukuda (University of British Columbia); Jongho Lee (University of British Columbia)*

The instability of supported liquid membranes (SLMs) and lack of monitoring tool for the membrane degradation remain a major hurdle to their use in resource recovery from waste streams. We present a novel approach of real-time, non-destructive monitoring of SLM degradation using electrochemical impedance spectroscopy (EIS).

We first prepared SLMs by immobilizing dodecane into a microporous PVDF substrate that is designed for selective recovery of medium-chain fatty acids (MCFAs) over short-chain fatty acids (SCFAs). The SLMs were modelled as a parallel electrical circuit of resistor (R) and capacitor (C), representing the defective and intact pores, respectively. We then employed four-electrode mode EIS to derive the values of R and C components over the course of SLM degradation. Four distinct phases of SLM degradation were identified, reflecting the varying rates of effective thickness and area reduction of the SLM. To validate this method, we tested SLMs at three different degradation stages: (i) intact, (ii) partially, and (iii) completely degraded SLMs as determined by EIS. The intact SLMs had exceptionally high selectivity (>200) for MCFAs over SCFAs. However, this selectivity drastically reduced to approximately 2.6 in the partially degraded SLMs and fell below 1 in the completely degraded SLMs, highlighting the significant impact of even minor degradation on the separation performance. The effective membrane thickness and area determined from the EIS measurements were then incorporated into a mass transfer model. The EIS-assisted model showed excellent agreements with the experimentally measured recovery and selectivity of SLMs at the three different stages of degradation. Our work highlights the efficacy of EIS for real-time monitoring of SLM structural changes and performance estimation, ensuring efficient SLM operations and stable SLM development.

2:40 PM

Oral 122 - Surface Modification of Nanofiltration Membranes for Selective Nutrient Recovery

Km Prottoy Shariar Piash (West Virginia University)*; Lian-Shin Lin (West Virginia University); Oishi Sanyal (West Virginia University)

This work focuses on the fabrication of polyelectrolyte-based nanofiltration membranes to selectively recover nutrients from nutrient-rich resources. Effluents of various wastewater sources contain substantial quantities of ammonium (NH4+) (300-700 mg/L) and potassium (K+) (13-120 mg/L)-based nutrients. These nutrients could serve as raw materials for fertilizer production and compared to traditional fertilizing raw material production processes, recovering nutrients from wastes present a more sustainable option. Furthermore, this also helps to avoid excessive nutrient buildup in lakes and rivers, which otherwise costs billions of dollars for remediation. In addition to valuable nutrients, effluents from concentrated wastewater streams also contain a significant amount of organic pollutants (TOC) (300-700 mg/L), whose presence can enable growth of micro-organisms and deplete oxygen supplies for aquatic life in freshwater sources. In this work, we focus on understanding the mechanism of NH4+, K+ and organic co- transport through NF membrane to achieve high NH4+ - K+/TOC selectivity at neutral pH. To accomplish this goal, a commercial NF membrane was utilized as support, and its selective layer was modified via layer-by-layer (LbL) deposition of two oppositely charged polyelectrolytes. This modification involved adjusting various LBL parameters, including polymer types, pH, the addition of salt (NaCl) within polyelectrolytes, polymer cross-linking, etc. Such modifications resulted in membranes with a range of properties, clearly demonstrating a trade-off between nutrient permeance and nutrient/organic selectivity. In addition, membranes with covalent crosslinking also facilitated intra-nutrient (NH4+/K+) selectivity, which is a rather surprising result, given that the two ions have exactly similar hydrated radius. Thus, this surface modification approach enables a final product with tunable ratios of nutrients, suitable for a diverse range of fertilizing applications.

Biochemical Applications

1:00 PM

Oral 123 - Membrane Optimization and Filtration Strategies Using Nanoparticles for Enhanced Efficiency in Gene Therapy Applications

Mara Leach (University of Kentukey)*; Gosia Chwatko (University of Kentucky); Dibakar Bhattacharyya (U. Kentucky)

With the escalating demand for adeno-associated viruses (AAVs) in gene therapy applications, the necessity for highly efficient purification methods has become imperative. This research concentrated on the pivotal role of membrane-based clarification in AAV harvest, utilizing silica nanoparticles as surrogates for a comprehensive exploration of membrane morphology. The importance of this investigation is highlighted by the inherent challenges of conventional purification methods, such as centrifugation, emphasizing the crucial requirement for scalable, efficient, and pristine AAV production. Employing nanoparticles as surrogates facilitated a systematic inquiry into the influence of membrane morphology on recovery rates. Specifically, 20nm nanoparticles emulate both the size and characteristics of AAVs, while larger particles (200nm) act as proxies for small cell debris, and BSA serves as a representative model for host cell proteins. Notably, our research revealed that a hydrophilic, high-porosity, asymmetric polyether sulfone (PES) membrane achieves ~100% recovery of 20nm particles, emphasizing the pivotal role of membrane selection. Surface fouling emerged as a critical concern during dead-end filtration, prompting an exploration of tangential flow filtration (TFF) as a strategic solution. Our findings indicated that TFF not only mitigates surface fouling but also provides a promising avenue for seamless integration into continuous manufacturing platforms, effectively addressing scalability issues. By addressing current limitations and proposing innovative filtration strategies, we lay the groundwork for the development of advanced and efficient AAV purification processes. This research was supported by the NSF EPSCoR Track 2.

1:20 PM

Oral 124 - Combined Hydrocyclone-Tangential Flow Microfiltration for Bioreactor Harvesting

Amir Hossein Mostafavi (University of Arkansas)*; Liang-Kai Chu (Pennsylvania State University); Xianghong Qian (University of Arkansas); Andrew Zydney (Pennsylvania State University); Ranil Wickramasinghe (University of Arkansas)

There is a growing interest in developing continuous biomanufacturing platforms. As opposed to a batch process, a continuous process runs at a steady state, which will lead to better control of product quality as well as the elimination of batch variation. In addition, more efficient use of process equipment will lead to a lower cost and more sustainable manufacturing process. Although perfusion bioreactors are

established for continuous biomanufacturing, the bioreactor clarification step is still challenging. While tangential flow filtration has many advantages in a continuous process, such as the direct introduction of the feed to the next capture step, fouling of the membrane, especially at the high cell titers (30-50 x 106 cells per mL or higher) used in perfusion bioreactors is a challenge. One way to avoid membrane fouling would be to lower the cell titer in the feed stream. Here, we have investigated an integrated hydrocyclonetangential flow filtration (HC-TFF) system. The hydrocyclone is used to produce two product streams: an underflow with increased particle concentration and an overflow that is depleted of particulate matter. To examine the hydrocyclone, Chinese hamster ovary (CHO) cells at different concentrations of 9 × 105 to 107 cells/mL were introduced to the hydrocyclone at various inlet velocities. The same overflow separation efficiency was obtained in the literature. Tests on the combined HC and TFF filter at a concentration of 3 × 106

cells/mL and low velocity of around 3.1 m/s showed that the cell viability remains about 97% after 2h. To improve the process design and gain more insight into the hydrocyclones, four hydrocyclones in a series were used and demonstrated a very high efficiency, which cannot be obtained with a single hydrocyclone. The outcome suggests that HC-TFF, if appropriately designed, has excellent potential for bioreactor clarification.

1:40 PM

Oral 125 - Charged biosponge polymers for capturing toxic chemotherapy drugs to minimize chemotherapy

Adele Godby (Pennsylvania State University); Rachel Lau (Pennsylvania State University); Hee Jeung Oh (Pennsylvania State University)*

Cancer is becoming the leading cause of death in most developed nations. Despite efforts to develop targeted and personalized cancer therapeutics, dosing of the cancer chemotherapeutics is limited by toxic side effects. Typically, more than 90% of the injected drug is not trapped in the target organ and bypasses the tumor to general circulation, causing toxicities in distant locations. In the context of reducing the toxicity of chemotherapy, we have designed charged, ion-exchange biosponge adsorbers for capturing toxic ionic chemotherapy drugs before they spread through the body. Doxorubicin, a widely used chemotherapy drug with significant toxic side effects, is a model drug. Charged nanostructured block copolymers with varied charge concentration are used to achieve high binding capacity and fast binding kinetics. The middle block is polystyrenesulfonate which has a high affinity for doxorubicin. We aim to understand the mechanism of ionic drug transport in these biosponge adsorbers. Using a swine model, our initial design enables the capture of 69 % of the administered drug without any adverse effects. The proposed approach may help patients fight cancer.

2:00 PM

Oral 126 - Mechanisms of Flux Decline in Crossflow Ultrafiltration of Whey Protein: Comparing Different Models to Experimental Data

Burcu Ozdemir (UCLA)*; Eric Hoek (UCLA)

Blatt et al.(1970) presented the first crossflow ultrafiltration model of protein concentration polarization (CP) and gel formation, which they adapted from Sherwood et al.(1965). Porter (1972) noted the underprediction of limiting fluxes from the model and postulated various hydrodynamic back-migration mechanisms to explain the discrepancies. Trettin&Doshi(1980) came up with a more fundamental derivation using an integral method solution to the general mass balance equation, which improved the model predictions. Song(1998) noted the integral method employs an assumed nth degree polynomial concentration profile to obtain a closed form solution, but Song then solved the general mass balance equation directly without an assumed concentration profile. Rabiller-Baudry(2000) incorporated an electrophoretic migration term into a film theory-based CP model (CDE model) with strong agreement between experimental data and model results. Here, we present crossflow ultrafiltration experimental data and mechanistic modeling of Lactalbumin and bovine serum albumin suspensions, which were prepared in solutions with total ionic strengths from 1-100 mM, and the experiments spanned a range of crossflow velocities (CF) and applied pressures up to the limiting flux. As predicted by all the gel-polarization models, the limiting flux increased with increasing CF and decreased with increasing protein concentration and ionic strength. The three classical gel-polarization models and the CDE model are compared with the experimental data and show varying fits. In this talk, we will discuss the experimental data and model fits for their abilities to accurately predict the limiting flux and membrane wall concentration. We also used the transient local model of Elimelech&Bhattacharjee(1998) to investigate the transition from CP only to gel layer formation, and the local variation of permeate flux along the length of the channel for the various solution and operating conditions.

2:20 PM

Oral 127 - Protocol for purification of circular RNA from a self-splicing reaction by ultrafiltration

Karen d Guillen Cuevas (Clemson University); Marc Birtwistle (Clemson University); Scott Husson (Clemson University)*

The development of messenger RNA (mRNA) vaccines against COVID-19 renewed the interest in using mRNA to treat existing and emerging diseases. One challenge for using mRNA, however, is the short duration of therapeutic protein expression due to its enzymatic degradation at the free ends. Circular RNA (circRNA) closed structure limits degradation by exonucleases, increasing its pharmaceutical stability compared to linear mRNA. Yet, the immunogenicity and protein expression stability of circRNA depend strongly on purity. The current purification method is high-performance liquid chromatography, which has a low recovery yield (about 50% for 90% purity) due to overlapping peaks. This method does not scale well and is not an option for commercial therapeutic circRNA purification. Thus, there is a pressing need for innovative, easy-to-implement, scalable purification methods to recover circRNA from reaction solutions with high purity and yield. The objective of this study is to investigate whether ultrafiltration (UF) can be used to purify protein-encoding circRNA from self-splicing reaction solutions with >95% purity and >70% yield. In this presentation, I will share results that quantify the critical fluxes of purified fractions of linear and circular RNA conformers and exploit differences in these critical fluxes to purify circRNA from binary mixtures. These data ultimately will be used to establish a protocol for purifying circRNA directly from self-splicing reaction solutions using UF, which is expected to increase purity and yield significantly and make it an attractive option for commercial circRNA purification.

2:40 PM

Oral 128 - Patterned anion exchange membrane for improved performance of glucose fuel cells

Thomas A DosSantos DiSorbo (University of Colorado Denver | Anschutz)*

We are developing a patterned anion exchange membrane (AEM) for abiotic glucose fuel cells (AGFC). AGFCs generate electrical power by catalyzing the oxidation of glucose to create an electrical current.

AEMs are an important component of AGFCs, facilitating selective ionic transfer between the cell cathode and anode. Many implantable medical device technologies require a safe and efficacious electrical power source to function over a prolonged period of use. To increase device longevity and reduce the complications of battery replacement, energy harvesting technologies like AGFCs are under investigation as viable biocompatible energy sources for medical devices.

We report benchtop testing of membrane electrode assemblies (MEAs) to study the advantages of using a patterned AEM in an abiotic, Pt-catalyzed fuel cell. Patterned AEMs have a larger surface area and should better facilitate ionic transport. The superior conductivity of patterned membranes have been observed in proton exchange membranes (PEMs) for hydrogen fuel cells, but the technology is novel to AEMs.

We evaluated AEMs that were either flat or imprinted with a nanometer-scale sinusoidal pattern across fuel mixtures containing various concentrations of KOH and glucose. The nanoscale AEM patterning yielded a statistically significant increase in open circuit voltage across all fuel ratios, indicating that an a priori overpotential mitigation mechanism exists for our MEAs. AEM patterning also increased maximum power density in trials with a low-alkalinity anode fuel. These results encourage further improvements in maximum power density and clinical utility for such fuel cell power supplies going forward.

Abstracts – Oral Presentations, Tuesday Evening

Presenting authors are indicated by an asterisk*.

Desalination and Osmotic Processes III

3:30 PM

Oral 129 - Permeance Prize: Thin Film Composite Membranes are not Just for Reverse Osmosis Anymore: Recoupling Materials Discovery with Thin-Film Manufacturability

Jeffrey R McCutcheon (University of Connecticut)*

For decades, a small group of membrane manufacturing processes has been used to make a vast majority of membranes used in separations Technology. In broad terms, casting, extrusion, and in-situ polymerization are the dominant manufacturing platforms currently used. This relatively small toolbox for membrane manufacturing has benefits and drawbacks. Some benefits include the general availability of the tools along with individuals with decades of experience in using them. The drawbacks are that these approaches may not be appropriate for new materials being developed for membrane technology. In fact, one of the reasons that new materials have not been adopted into commercial membranes may be that they are not easily processable into membranes using these tools. It is important that we adopt new membrane manufacturing technologies that enable the use of new materials. Our group has pioneered one such technology, electrospray deposition, that enables control of morphology, chemistry, and thickness. We use this technology for making thin film composite (TFC) membranes from conventional and unconventional polymer materials for use in water treatment, gas separations, and electromembrane applications. As a mostly material agnostic processing technology, it can enable the TFC membrane platform, and all of its benefits, for applications outside of reverse osmosis and nanofiltration.

3:50 PM

Oral 130 - Evaluation of Commercial Nanofiltration and Reverse Osmosis Membrane Filtration to Remove PFASs: Effects of Transmembrane Pressures, Water Matrices such as Surfactants, and Ion Valency and Temperatures

Guangyu Zhu (New Jersey Inst of Technology)*; Boris Khusid (New Jersey Inst of Technology); Wen Zhang (New Jersey Institute of Technology); Qingquan Ma (New Jersey Inst of Technology); Neel Ahuja (Millburn High School)

Per- and polyfluoroalkyl substances (PFAS) have emerged as an environmental concern due to their persistence and potential health risks. Among different PFAS treatment or separation technologies, nano-filtration (NF) and reverse osmosis (RO) both yield high rejection efficiencies of diverse PFAS in water. However, both technologies are affected by many intrinsic and extrinsic factors. This study evaluated the rejection of PFAS of different carbon chain length by two commercial RO and NF membranes under different operational conditions (e.g., applied pressure and initial PFAS concentration) and feed solution matrixes, such as pH (4~10), salinity (0~1000 mM NaCl), and organic matters (0~10 mM). Moreover, this study investigated the influence of solution temperature, surfactants, and ion valency on the rejection of PFOA and PFBA. Our results confirmed that size exclusion is a primary mechanism of PFAS rejection by RO and NF, as well as the fact that electrostatic interactions are important when PFAS molecules have sizes less than the NF membrane pores. The results reveal that temperature variations significantly affect PFAS rejection efficiency, which decreased appreciably at high solution temperatures primarily because an increase in temperature leads to thermally expanded RO/NF membrane pores, allowing for the passage of additional PFAS molecules. The presence of cationic surfactants increases PFAS rejection by binding with negatively charged PFAS molecules and preventing them from passing through membrane pores via size exclusion. Moreover, Al3+ and PO43− with higher ion valency enabled higher rejection of PFOA and. Overall, this research contributes valuable insights into the factors influencing PFAS rejection in commercial RO and NF processes.

4:10 PM

Oral 131 - A Concise Analytical Model for the System Performance of an RO Desalination Process

Lianfa Song (Texas Tech University)*

Permeate recovery is a key parameter that plays a central role in the performance assessment and optimization of reverse osmosis processes. It remains a great challenge for engineers in the field to determine the recovery conveniently and accurately from the basic parameters of a membrane system. A concise analytical model is presented here that, without the need for the empirical or fitting coefficients and tedious numerical calculation, links the recovery of a reverse osmosis process rigorously to three basic parameters: the feed water salt concentration, the characteristic of membrane process, and the driving pressure. A graphical solution method is also introduced to find the recovery more effortlessly. The optimization of an RO desalination process can be greatly streamlined when concise model is used to calculate the recovery of the membrane system. The impacts of energy, membrane, feed water pretreatment, and concentrate handling on the optimal recovery of a membrane desalination process are demonstrated. The avoidance of a large number of equations and numerical integration in the determination of permeate recovery can greatly facilitate the optimization of design and operation of seawater desalination processes.

4:30 PM

Oral 132 - Novel artificial water channel membranes with improved molecular separation properties in reverse osmosis

Maria Di Vincenzo (KAUST)*; Suzana Nunes (King Abdullah University of Science and Technology (KAUST)); Alberto Tiraferri (Politecnico di Torino)

The fundamental understanding of supramolecular water channels characterized by sub-nanometer assemblies has made biomimetic approaches to developing polymeric membranes for desalination applications increasingly attractive. Their separation efficiency lies in controlling the selectivity of water at the molecular level, enabling the possibility to overcome the perm-selectivity trade-off of traditional thin-film composite (TFC) polyamide membranes.

In our previous collaboration, enhanced reverse osmosis (RO) membranes based on imidazole quartet artificial water channels (AWC) were demonstrated as a very promising and competitive approach for desalination [1]. We are now exploring a novel AWC structure that self-organizes and could be seamlessly integrated within polyamide active layers during a streamlined interfacial polymerization procedure. The resulting defect-free layers were designed to enhance the removal of neutral solutes and improve the perm-selectivity performance of traditional TFC membranes. The most advanced physicochemical characterization techniques (i.e., cryo-TEM, GIWAXS, XPS) were used to delve into the kinetic formation of the new bioinspired layers, unveiling their nanostructural features.

Experimental results demonstrate that our membranes can provide selective water transport and enable exceptional Å-scale separations. Their performance has been assessed by cross-flow filtration under representative brackish water and seawater desalination conditions at 15 bar and 55 bar of applied pressure, respectively. As a result, bioinspired membranes outperform commercial state-of-the-art and control TFC membranes by 250% in terms of water permeance, while maintaining excellent observed NaCl rejection (99.5%). Ultimately, they display better selectivity toward neutral solutes such as urea and boron (up to 20% higher), compared to the best-marketed seawater RO membranes (e.g. HRLE SW30 Film Tec).

[1] Nature Nanotechnology 16.2 (2021): 190-196

4:50 PM

Oral 133 - Regulating interfacial polymerization via multi-functional calcium carbonate based interlayer for highly permselective nanofiltration membrane

Mengyang Hu (Kobe University)*; Hideto Matsuyama (Kobe University)

Highly permselective nanofiltration (NF) membranes have potential in water treatment applications, but overcoming the permeability-selectivity trade-off is a challenge. We introduce in this work a highly permselective NF membrane fabricated by regulating interfacial polymerization (IP) with a calcium carbonate (CaCO3) interlayer. The flexible CaCO3 interlayer was grown in situ onto the surface of polyketone (PK) membrane through mineralization, which serves as a regulator to mediate the kinetics of IP by inhibiting the amine monomer diffusion across the interface between the aqueous and organic phases, promoting the synthesis of a thinner selective layer. Furthermore, the CaCO3 interlayer can react with HCl, a byproduct of the IP reaction, to release more heat and CO2 nanobubbles, resulting in the formation of a rough selective layer with loose structure. The resultant CaCO3 interlayer-incorporated NF membrane demonstrated a high water permeance of 23.4 L m-2 h-1 bar-1 (around 3.5 times the permeance of the pristine NF membrane (6.7 L m-2 h-1 bar-1)) with Na2SO4 rejection of 97.7%. The multi-functional interlayer explored in this current work may provide a new strategy to preparation of membranes with highly satisfactory NF performance.

5:10 PM

Oral 134 - Three-dimensional modeling of flow and mass transfer in spacer-filled channels in Reverse osmosis/Nanofiltration processes

Santiago Cespedes (KAUST)*; Fernan David Martinez Jimenez (KAUST); Bastiaan Blankert (KAUST); Cristian Picioreanu (KAUST)

The assessment of the mass transfer phenomena in processes such as Reverse Osmosis (RO) and Nanofiltration (NF) is fundamental for design and performance improvement. The mass transfer coefficient is mostly evaluated through correlations involving Reynolds and Schmidt numbers. However, these correlations give global values, mostly overlooking local phenomena such as quasi-stagnant flow caused by the presence of feed spacer. Thus, we need correlations for the local mass transfer that can be used in simplified design models (e.g. 1D).

We computed local and global mass transfer coefficients with a 3D numerical model for spacer-filled feed RO and NF channels by coupling hydrodynamics with the transport of solutes by convection and diffusion, and permeation through the membrane. A realistic spacer geometry was obtained by CT-scans of commercial spacers. We also compared the effect of different membrane lengths, from 1 cm to 1 m, on the global mass transfer coefficient. A length of 20 cm, in typical RO operation conditions, was found sufficient to achieve a relatively constant overall mass transfer coefficient.

The model shows a highly non-uniform distribution of concentration and mass transfer coefficients on the membrane (see figure). By systematically changing the design and operation variables (e.g., water velocity, pressure, salt concentration, membrane permeabilities, channel width, spacer geometry and orientation) we developed engineering correlations. These were critically compared with results from the literature, showing differences due to the presence of spacers. This model can be used to study the full length of an RO or NF membrane module and thus provide better indicators of the design and optimization of the separation process.

Keywords: Desalination; Membrane spacers; Computational fluid dynamics; Spacer geometry; Mass transfer.

NAMS Keynote I

3:30 PM

Oral 135 - A Tale of Two Carbon-Capture Properties: Bulk and Thin Films of Mixed Matrix Materials

Haiqing Lin (The State University of New York at Buffalo)*; Gengyi Zhang (The State University of New York at Buffalo); Lingxiang Zhu (National Energy Technology Laboratory)

Mixed matrix materials (MMMs) integrating excellent processability from polymers and distinct separation properties from nanofillers are of interest for membrane gas separations. However, they are difficult to fabricate into thin-film nanocomposite (TFN) membranes due to interfacial incompatibility between polymers and nanofillers. Here we have successfully developed TFN membranes based on MMMs (as thin as 200 nm) comprising amorphous poly(ethylene oxide) (aPEO) and UiO-66-NH2 enabled by the strong hydrogen bonds between the two matrices. Increasing the UiO-66-NH2 loading unexpectedly decreases CO2 permeability in freestanding films (200 – 300 μm), but it surprisingly leads to the best CO2/N2 separation properties in the membranes at a loading of 10 mass% (CO2 permeance of 2900 GPU and CO2/N2 selectivity of 48). Nanoconfinement exerts a significant influence on the morphological and gas separation properties of the MMM layer. The membrane with 10 mass% UiO-66-NH2 demonstrates mixed-gas CO2 permeance of 1400 GPU and CO2/N2 selectivity of 76 in the presence of 1.2 mol% water vapor at ≈23 °C, surpassing Robeson's upper bound. The membrane also demonstrated stable CO2/N2 separation performance when challenged with real flue gas for 700 h continuously at National Carbon Capture Center (NCCC).

4:00 PM

Oral 136 - Osmoses: Developing Gas Separation Membranes to Enable Sustainable Energy Solutions

Holden W Lai (Osmoses)*

From a hydrogen economy to low-carbon molecule alternatives, the inefficiency of existing gas separation technologies remains a bottleneck to the implementation of clean energy solutions and the ultimate goal of decarbonizing and hitting net zero. The norbornyl benzocyclobutene polymer platform, first developed by the Xia lab at Stanford University, has great potential as materials for highly efficient membrane gas separations (Liu et al. J. Am. Chem. Soc. 2014). Norbornyl benzocyclobutene ladder polymers have demonstrated thermal, mechanical, and chemical stability, and their bulk films have shown exceptional combinations of permeability and selectivity after aging (Lai et al. Science 2022). Founded in 2021, Osmoses Inc. took on the challenge to realize the exceptional potential of norbornyl benzocyclobutene membranes in commercial gas separation systems. We redesigned the polymer chemistry to facilitate scale up, developed and fabricated thin-film composites to improve permeance and selectivity, and assembled membrane modules to prepare for field trials. We are now working with end users and partners to pilot our technology, with the ultimate goal of implementing it broadly and rebuilding industrial infrastructure so that separations are never again a bottleneck to a cleaner world.

4:30 PM

Oral 137 - Deep Flue Gas Decarbonization with Novel Membrane Process Designs

Tim Merkel (Membrane Technology & Research Inc.); Trevor Carlisle (Membrane Technology & Research Inc.); Pingjiao Hao (Membrane Technology & Research Inc.); Richard Baker (Membrane Technology & Research Inc.)*

CO2 capture and sequestration using membranes is a potentially cost-effective way to reduce global emissions from power plants, refineries, cement and steel plants. The target of the separation process is capture of ~90% of the CO2 in the feed gas into a permeate stream containing 99% CO2. A single stage membrane system can achieve 90% capture of CO2 or 99% purity of CO2 but not at the same time. Better, more selective membranes are always helpful, but the problem cannot be solved by an improved membrane alone. To obtain a complete separation a number of membrane separation units must be combined into an integrated flow scheme.

Cost effective high CO2 capture and concentration is particularly difficult from dilute flue gas streams such as those produced by gas turbines, biomass to energy power plants and steam boilers. One novel way to treat these gas streams is to use a newly developed process called Selective Exhaust Gas Recycle (SEGR). SEGR uses a membrane contactor to selectively remove residual amounts of CO2 from the flue gas into a sweep air stream. The sweep air containing some CO2 is used in the combustion process. The result is to obtain high CO2 capture at a low energy cost.

The DOE has sponsored membrane process development programs at MTR for over a decade. The technology has advanced from laboratory feasibility studies to the construction of a 100 ton/day CO2 capture pilot plant at a coal power station. In this talk the factors that went into the design of this capture plant and into other capture applications where SEGR is used will be discussed.

5:00 PM

Oral 138 - Polymer membranes stability under sour-humid natural gas mixtures and their role towards de-carbonization

Nitesh Bhuwania (Chevron)*; Daniel Chinn (Chevron); Thijs Peters (SINTEF); Luca Ansaloni (SINTEF); Oguz Karvan (EMI Twente BV); Alberto Tena (EMI Twente BV); Tymen Visser (EMI Twente BV)

For industry to adopt polymer membranes, it is crucial to understand the separation behavior of the membranes under relevant field conditions. The understanding of the influence of high levels of hydrogen sulfide (H2S), moisture and other constituents on the polymer membrane separation performance and stability under these conditions is very limited. This talk will highlight the separation performance of various membrane materials in H2S-containing mixtures (up to 15 mol% H2S) including various levels of humidity and in the presence of higher hydrocarbons, such as butane and toluene [1]. Based on different feed-gas test conditions, the separation performance and the stability of the investigated membranes are promising. The combination of H2S plasticization and competitive sorption in the presence of humidity indicates that membrane technology is attractive for humid natural gas application with high H2S content [1].

In addition, modular membranes can be integrated with the rest of the gas treatment plant for de-bottlenecking the plant and to reduce the carbon footprint by diverting the H2S and CO2 molecules back to the reservoir through reinjection [2]. We will present a high-level understanding of how membrane technology can help in the carbon reduction journey in such applications.

References

[1] Peters, T.A., et al., J. Membr. Sci., 2024, 693, 122324.
[2] Bhuwania et al., US Patent 10 (905) (2021) 996, Chevron USA Inc.

2D Membrane Materials and Processes

3:30 PM

Oral 139 - Strengthening Isoporous Polymer Membranes through Chain Entanglements

Adam N Mann (University of Texas at Austin)*; Benny Freeman (University of Texas at Austin); Gabriel Sanoja (University of Texas at Austin)

Ultrafiltration (UF) membranes find widespread use in water purification and bioprocessing. However, it remains challenging to rationally co-design their mechanical and transport properties. This is for two reasons: (i) the typical fabrication process, nonsolvent-induced phase separation, kinetically arrests a structure with broad pore size distributions on the surface and in the bulk and (ii) these pores play a crucial and contrasting role on mechanical and transport properties. The use of amphiphilic block copolymers offers narrow surface pore size distributions by taking advantage of their solution self-assembly. However, most of these block copolymers are based on polystyrene, a notoriously weak and brittle polymer, due to a high packing length which reduces its ability to entangle. Engineering strong block copolymers can solve the first challenge and circumvent the second, thereby enabling the development of the next generation of advanced UF membranes.

Here, we introduce poly[(styrene-co-acrylonitrile)-block-4-vinyl pyridine] (SAN-P4VP) as a platform for designing permeable, selective, and mechanically strong membranes. By considering block copolymers of constant molecular weight, Mn ≈ 100 kDa, and acrylonitrile content from 0 to 40 wt%, we demonstrate three important points. The first is that incorporating acrylonitrile groups in the main chain of the hydrophobic block strengthens the polymer membranes by decreasing the average packing length along the chains and enabling entanglements. The second is that the narrow pore size distributions and pure water permeance values of the membranes are unaffected by the presence of acrylonitrile. And third is that these block copolymers are easy to synthesize via a RAFT emulsion copolymerization of acrylonitrile and styrene using poly(4-vinyl pyridine) as a macromolecular chain transfer agent. These results pave the way for developing advanced UF membranes and improving separations in healthcare, energy, and water.

3:50 PM

Oral 140 - Highly Selective Thin B-Oriented MFI Zeolite Membranes on Scalable Modified Stainless-Steel Supports

Fateme Banihashemi (Arizona State University); Jerry Lin (Arizona State University)*

Enhancing the separation performance of molecular sieve membranes relies significantly on precise control over their membrane morphology and microstructure. This involves creating thin, closely packed, oriented, and uniform coatings of seed crystals, enabling subsequent intergrowth to establish continuous membranes. To address industrial needs, it is crucial to generate a thin and flawless membrane efficiently producible on a large-scale support. Porous stainless supports offer scalability and easy integration into membrane modules. In this study, we present a novel and straightforward process for modifying stainless-steel supports to seed zeolite membrane layers. High-performance, b-oriented MFI zeolite membranes can be synthesized on these modified stainless-steel supports through scalable filtration seeding of MFI zeolite nanosheets followed by secondary growth. By employing a carefully designed pre-treatment procedure, we achieved the growth of well-intergrown, highly b-oriented MFI zeolite membranes on the modified stainless steel support, establishing a robust interaction with the substrate. The membranes, fabricated through gel-free secondary growth of nanosheets deposited on a scalable stainless steel support using this technique, demonstrate exceptional ultra-selective capabilities in separating para-xylene from ortho-xylene, with a high separation factor of approximately 400.

4:10 PM

Oral 141 - Ion-sieving membranes based on Polymers of Intrinsic Microporosity for aqueous organic redox flow batteries

Chunchun Ye (School of Chemistry, University of Edinburgh)*; Neil McKeown (School of Chemistry, University of Edinburgh); Qilei Song (Department of Chemical Engineering, Imperial College London); Anqi Wang (Department of Chemical Engineering, Imperial College London); Rui Tan (Warwick Electrochemical Engineering, Warwick Manufacture Group, University of Warwick)

Redox flow batteries (RFBs) using aqueous organic-based electrolytes are promising candidates for cost-effective grid-scale energy storage. The membrane separator is a key component determining RFB performance, however, designing separators that facilitate fast transport of charge carrier ions while preventing the crossover of redox-active molecules remains challenging.

Here, we report the development of easily processed ion-sieving membranes based on Polymers of Intrinsic Microporosity (PIMs) and demonstrate their promising performance in aqueous organic RFBs. Our design approach combines 1) the use of PIMs with defining features of rigidity and contorted backbones to generate a large amount of microporosity with a narrow pore size distribution; 2) the introduction of hydrophilic functionalities to facilitate the formation of sub-nanometre-sized water channels for ion conduction. The micropore architecture of PIM membranes can be finely tuned by adjusting the rigidity of structural units, and the micropore chemistry can be controlled by introducing different types of functional groups while modifying their contents. These PIMs can be processed into robust self-standing membranes via simple solution processing. As-prepared PIM membranes demonstrate the critical role of precise control over both pore architecture and chemistry in achieving rapid and selective ion transport.

Two emerging aqueous organic cell chemistries for grid RFBs are applied to uncover the relationship between membrane properties relevant to cell efficiency and cycle life. Importantly, the new PIM membranes significantly enhance battery energy efficiency and peak power density while enabling stable operations in aqueous electrolytes, promising long-lasting, crossover-free RFBs. The confined ion transport at the sub-nanometre scale in PIM membranes suggests their potential utility for various energy generation and storage devices, as well as water purification.

4:30 PM

Oral 142 - Role of salt additives in tuning the structure of β-cyclodextrin-based membranes

Jinting Hu (KU Leuven)*; Sem Bleus (KU Leuven); Linde Achten (KU Leuven); Wim Dehaen (KU Leuven); Xing Yang (KU Leuven)

Macrocycle molecules with intrinsic nanoscale cavity have great potential in constructing ultra-permeable thin-film composite (TFC) membrane with permanent porosities to overcome the inherent permeability-selectivity trade-off. Efforts to optimize the interconnection of the cavity-converted porosities focus on regulating macrocycle monomer diffusion and polymer growth in the interfacial polymerization (IP) process.

This study explores the impact of common inorganic salt additives in IP layer formation of β-cyclodextrin-based TFC membranes. The TFC membranes were synthesized on a polyethersulfone substrate made by phase inversion. β-cyclodextrin (β-CD, 4 wt%) dissolved in NaOH (2.0 wt%) aqueous solution was crosslinked with trimesoyl chloride – n-hexane solution (TMC, 0.2 wt% and 1.0 wt%) to create the polyester top layer. The addition of salts (LiCl, NaCl, KCl, and K2SO4, 3.0 wt%) in the aqueous solution during IP resulted in increased water permeance, e.g., the KCl-mediated membrane achieving an ultra-high permeance of ~691.4 L·m-2·h-1·bar-1 without compromising selectivity, which has never been reported by other β-CD TFC membrane research. The effects were attributed to the β-CD diffusivity changes in salt-assisted IP environment, leading to tunable nanoscale homogeneities and reduced channel blockages. Also, salt crystals precipitated in the interface during heat curing, creating extra free voids and cavity interconnections. Moreover, the salt additives decreased the specific heat capacity of aqueous solutions, triggering fiercer Rayleigh-Bénard convection flow and shaping a more crumpled IP film. The salt-mediated membranes demonstrated stability and efficient solute rejection of selected cationic and anionic dyes, making them promising for applications such as dye recycling from textile wastewater with high salinity. Overall, this research provides green route to enhance the performance of macrocycle-based membranes through salt-mediated IP method.

4:50 PM

Oral 143 - Revealing Molecular Mechanisms in Hierarchical Nanoporous Carbon via Nuclear Magnetic Resonance

Mao, Haiyan *; Halat, David; Pines, Alexander; Cui, Yi; Rimer, Jeffrey

Hierarchical nanoporous carbons (HNC) have been proven to be an effective adsorbent for the adsorption of volatile organic compounds (VOCs) and CO2, although questions remain regarding the hierarchical structure regulation, the adsorption mechanisms of adsorbate uptake and interactions within the HNC. We synthesize HNC from wood, using a microwave-induced heating method incorporating K2CO3 activation. HNC exhibit Murray's Law multi-scale structures, prompting a molecular scale study of adsorbate adsorption via nuclear magnetic resonance (NMR). NMR chemical shifts are consistent with ring current effects from the adsorbent. Our NMR technique provides a convenient way to quantitate adsorption of adsorbate in HNC. VOCs vapor adsorption results show NMR chemical shift changes with time, suggesting initial adsorption into mesopores, followed by diffusion into micropores. Schroeder's Paradox is demonstrated by differences in observed shifts for adsorbed liquid vis-à-vis vapor phase in these HNC. These HNC show high CO2 adsorption capacity, portending applications to carbon capture.

References:

[1] Mao, H. et al. Alexander Pines, Yi Cui and Jeffrey A. Reimer. Matter 2020, 3, 2093-2107.

[2] F.K. Yuen, B. H. H., Adv. Colloid Interfac. 2009, 19-27.

[3] D. Hernández-Monje, et al. Molecules 2018, 23, 476.

[4] C. Merlet, A. C. Forse, J.Griffin, D. Frenkel, and C.P. Grey, J. Chem. Phys. 2015, 142, 094701-094713.

Acknowledgements: This work was partly supported by the DOE, Office of Basic Energy Sciences, Division of Materials Sciences and Engineering (contract no. DE-AC02-76SF00515). J.A.R. acknowledges partial support from the ACT-PrISMa project, which has received joint funding from BEIS, NERC, and EPSRC (UK), funding from the Division of CCS R&D, DOE, and funding from the Office Fédéral de l'Energie (Switzerland)

5:10 PM

Oral 144 - Anti- swelling Characteristics Enhanced via Self-Cross-Linking in MXene-Interlaced Graphene Oxide Membranes

Saurabh Kr Tiwary (University Of Houston)*; Alamgir Karim (University of Houston)

Membrane-based water purification holds immense promise in addressing the global challenge of ensuring safe and clean water access. Numerous studies have concentrated on developing nearly two-dimensional membranes for this purpose, employing an ion rejection technique. However, a significant hurdle in these materials lies in membrane swelling, jeopardizing their functionality, and using the proposed graphene oxide (GO) stack for desalination results in poor outcomes due to GO's high-water affinity. Immersing the stack in water causes two to three layers of water molecules to intercalate between pores, leading to swelling and increased pore size. This allows salt atoms to permeate. To counter this, GO laminates are constrained from swelling using Self-Crosslinking of MXene-intercalated GO Membranes, MXene nanosheets, and GO alternate to form filtrating membranes, addressing the swelling issue in desalination. Presenting a breakthrough in addressing this critical issue, our study introduces a self-cross-linked MXene-intercalated GO membrane. This innovative membrane maintains its ion and dye rejection capabilities due to the robust physical cross-linking interaction between Ti–O–Ti bonds and neighboring nanosheets, effectively curbing membrane swelling. Additionally, the formation of associative bonds such as C–O–C, O=C–O, and C–OH further contributes to the inhibition of membrane swelling. To ensure the sustained performance of these membranes in practical applications, the resulting membrane demonstrates minimal swelling when immersed in solutions, ensuring the retention of its functionality. Even when challenged with varying concentrations of salt and dye, the membrane consistently rejects these contaminants over an extended period. This longevity suggests that the proposed membranes possess the requisite durability for water filtration applications, marking a significant advancement in membrane technology focusing on anti-swelling properties.

Viral and Antibody Filtration

3:30 PM

Oral 145 - The application of tangential flow filtration for the continuous processing of monoclonal antibodies: process optimization based on particle morphology

Ali Behboudi (The Pennsylvania State University)*; Andrew Zydney (Pennsylvania State University)

Recent developments in cell culture have resulted in significant increases in monoclonal antibody titer, which has led to a resurgence of interest in the utilization of precipitation for the initial capture and purification of high-value therapeutic proteins. The precipitated protein can then be dewatered and washed by tangential flow microfiltration using hollow fiber membrane modules. One of the key challenges that must be overcome to develop a continuous precipitation-filtration process is the need to eliminate membrane fouling. The objective of this work was to examine the possibility of controlling the particle morphology to improve the filterability of the precipitated particles. Zinc chloride and polyethylene glycol were used as precipitating agents for human serum immunoglobulin G (IgG), both with and without added salts. Incorporating 300 mM CaCl2 or Na2SO4 in the precipitation solution enhance the performance of the TFF process, increasing the critical flux by 25%. Further improvements in the critical flux were obtained using Dibasic Sodium Malonate (DSM) at concentrations as low as 20 mM. This increase in the critical flux was not merely the result of a change in the particle size distribution; rather, it was related to an increase in the density of the precipitate. Further investigations revealed that this change in particle morphology and consequently, precipitate filterability, was due to a reduction in the precipitation kinetics caused by the different salts. Additional insights were obtained using a variety of biophysical characterizations including DSC and FTIR Spectroscopy. In addition, the precipitate density could be rapidly measured by centrifugation providing a practical, easy-to-calculate, parameter to identify precipitation conditions that lead to good filterability. During a 24-hour filtration process minimal fouling with high degrees of purification, proved the viability of this process for the purification of monoclonal antibodies

3:50 PM

Oral 146 - A more sustainable all-membrane process for the purification of monoclonal antibodies

Cristiana Boi (Università di Bologna)*; Jinxin Fan (North Carolina State University); Joseph Lavoie (North Carolina State University); Ruben Carbonell (North Carolina State University)

The biopharmaceutical industry uses significant amounts of water and energy in processes such as fermentation, mammalian cell growth, chromatography, filtration, and viral inactivation. While this industry is not a major contributor to climate change, there is increasing interest in reducing environmental impact by developing more efficient processes that could improve the sustainability and efficiency of biologics manufacturing.

In this context, membranes play an important role in downstream processing, enabling higher productivity while reducing buffer usage, production costs and process mass intensity. Chromatographic membranes with high binding capacity can be operated with much shorter residence times than columns, resulting in more than a 10-fold increase in productivity.

To demonstrate these concepts, we have developed a non-protein A process for the purification of a mAb from clarified cell culture broth. The process uses only nonwoven membranes and includes a capture step with a novel multimodal ligand with mild elution conditions, viral inactivation, a flow-through anion exchange purification step, and a final bind-and-elute cation exchange polishing and concentration step.

Using this process, we were able to reduce HCP from 327000 ppm in the supernatant to 93.5 ppm (3.5 LRV), DNA from 994.7 ppm to 6.8 ppb (5.3 LRV), % aggregates from 5.4% to 0.4% with no detectable fragments, thereby meeting the common requirements (HCP<100 ppm, DNA<10 ppb, % aggregates<1%) for mAb products. A side-by-side comparison with the standard mAb platform (protein A resin and two ion exchange resins) showed that the three-step membrane process outperformed the standard process with significantly shorter process time (3.8 h vs. 13.1 h), 14% higher overall recovery (88.3% vs. 77.5%), fewer steps (no diafiltration for buffer exchange) and fewer aggregates in the final product despite similar HCP and DNA removal.

4:10 PM

Oral 147 - IgG-BSA Separation and Purification by Internally Staged Ultrafiltration

Lixin Feng (New Jersey Institute of Technology); Yufeng Song (New Jersey Institute of Technology); Sagnik Basuray (New Jersey Institute of Technology); Kamalesh K Sirkar (New Jersey Institute of Technology)*

Purification of IgG from residual host cell proteins (HCPs) in post-Protein A chromatography is important since some HCPs bind with Protein A and elute with the mAb; removal of HCPs from CHO cell lines is essential. To that end, an advanced separation and purification technique in biopharmaceutical manufacturing, namely, internally staged ultrafiltration (ISUF), is investigated here. This builds on successful results obtained in separating a binary protein mixture of hemoglobin and bovine serum albumin (BSA) into two highly purified fractions by using ISUF with a stack of three identical ultrafiltration (UF) membranes. Choosing BSA as a model for HCPs, separation of a binary mixture of IgG and BSA containing 1.0 mg/ml IgG and 0.1mg/ml BSA is successfully demonstrated here using a modified ISUF technique: two Omega 100 kDa membranes on top followed by one Omega 70 kDa membrane at the bottom. This modified configuration demonstrated exceptional performance with almost 100% rejection, 99% purity, and 99.5% retention of IgG, along with 96.5% recovery of BSA over 10 diavolume. This modified membrane stacking resulted from strategic considerations of membrane stacking and careful selection of molecular weight cutoffs and materials, and performance analysis of different membrane stacking configurations using rejection behaviors, purity levels, and recovery rates under varying diavolumes and pressure differentials. This approach enhances flexibility in membrane choices in ISUF and provides valuable insights for optimizing membrane-based biopharmaceutical separation techniques.

4:30 PM

Oral 148 - Effect of protein on the location of virus entrapment during virus filtration

Wenbo Xu (University of Arkansas); Hironobu Shirataki (Asahi Kasei Corporation); Xianghong Qian (University of Arkansas); Ranil Wickramasinghe (University of Arkansas)*

"Validation of virus contamination is essential during the downstream purification of biopharmaceutical products. In the case of protein based therapeutics, virus filtration is routinely used to validate virus clearance. Adventitious virus clearance is frequently validated using minute virus of mice (MVM) a FDA recommended model parvovirus 18-22 nm in size. Virus filters are run in dead end mode. They are designed to achieve 4 or more log (10,000 fold) virus clearance in the permeate relative to the feed. In addition, more than 95% product recovery in the permeate is required. Often the product of interest may have a hydrodynamic radius in excess of 10 nm leading to a very challenging separation. In practice the performance of the virus filters has been shown to be affected by the product species.

In this study the effect of a monoclonal antibody (mAb) as well as BSA on the location of capture of MVM in three different parvovirus retentive filters has been studied. All experiments were run under constant flux conditions. The membranes were challenged with 1,000 L m-2 of feed. The concentration of aggregates for the mAb was less than 1% but 15 % (dimer and higher oligomers) for BSA. The MVM and BSA were labelled with a fluorescent dye. Laser scanning confocal microscopy was used to locate the position of virus retention on the membrane. The presence of mAb or BSA shifted the location of MVM retention toward the permeate side of the membrane. The effect was more significant for BSA. BSA aggregate retention occurred closer to permeate side of the membrane compared to MVM. The inclusion of a process pause also led to virus retention close to the permeate side of the membrane due to back diffusion of the virus particles. All three virus filter showed virus clearance in excess of 4 log under all conditions. BSA and mAb recovery was over 99%. The result indicate the effect of product aggregates as well as interactions between product aggregates and MVM on virus clearance.

4:50 PM

Oral 149 - The Effects of Pressure and Process Interruption on The Location of Virus Particles Captured during Virus Filtration

Wenbo Xu (University of Arkansas)*; Hironobu Shirataki (Asahi-Kasei); Ranil Wickramasinghe (University of Arkansas); Xianghong Qian (University of Arkansas)

Viral safety is one of the major concerns in the production of mammalian cell and plasma derived biotherapeutic products. A demonstration of virus clearance is required by the regulatory agencies. Virus filtration can provide a robust removal of virus. Virus filtration unit operation has been integrated into many downstream purification processes. Here the effects of pressure on the location of virus particles captured by three different virus filters are investigated using confocal imaging studies with labelled minute virus of mice (MVM). Our results indicate that reduced pressure leads to the migration of captured virus particles towards the permeate side during virus filtration. In addition, process interruption with a sudden drop of pressure has a significant effect on the locations where the virus particles are captured due to the back-diffusion and redistribution of the virus particles. Finally longer filtration time will also impact where virus particles are trapped in the filters. Our study has implications for virus filtration processes particularly during continuous biomanufacturing of therapeutic proteins.

5:10 PM

Oral 150 - Bioseparation Membranes from Crosslinking of Pluronic F127 Self-assembly

Tabatabaei, Seyed Mostafa (University of Oklahoma)*; Foudazi, Reza Mostafa (University of Oklahoma)

Ultrafiltration membranes are extensively used in various industries. One of the applications of ultrafiltration membranes is virus separation in monoclonal antibody (mAb) production. mAbs have a diameter of approximately 8-10 nm and parvoviruses are larger than 20 nm. Therefore, an ideal membrane should have a narrow pore size distribution with an average pore size between these ranges. Highly ordered structures can be obtained by using lyotropic liquid crystals (LLCs) as a template. LLCs are formed from the self-assembly of amphiphilic block copolymers in the presence of selective solvent(s) resulting in hierarchical mesostructures in the range of 2-50 nm. In this study, we investigate the self-assembly of Pluronic F127, an FDA approved surfactant, in ethylammonium nitrate, a hydrophilic ionic liquid. Successful discontinuous cubic structures are reported which can be modified in terms of both structure and domain size by varying the water content in the formulation. The retention of the structure with domain size of 20-26 nm after UV crosslinking of the LLC is confirmed which shows the potential of the system in membrane applications. The resulting membrane exhibits a hydrophilic surface with contact angle of 57°. The LLC templated membrane has a pore size of 18 nm with 3D liquid channel pathways and a narrow pore size distribution, which are appropriate for virus clearance in monoclonal antibody production.

Abstracts – Oral Presentations, Wednesday Morning

Presenting authors are indicated by an asterisk*.

Membranes for Carbon Capture II

8:30 AM
Oral 151 - Large pilot testing of MTR's membrane-based post-combustion CO2 capture process

Richard Baker (Membrane Technology & Research Inc.); Phillip Behm (Membrane Technology & Research Inc.); Alicia Breen (Membrane Technology & Research Inc.); Brice Freeman (Membrane Technology & Research Inc.); Pingjiao Hao (Membrane Technology & Research Inc.); Thomas Hofmann (Membrane Technology& Research Inc.); Jay Kniep (Membrane Technology & Research Inc.); Tim Merkel (Membrane Technology & Research Inc.); Witopo Salim (Membrane Technology & Research Inc.)*; Ray McKaskle (Trimeric Corporation); Dana Pierik (Sargent & Lundy); William Morris (Wyoming Integrated Test Center)

MTR is currently working to scale-up our capture technology as the final step of commercialization. In this presentation, we will describe our current Department of Energy (DOE) sponsored Large Pilot project which encompasses the design, build, and operation of a 150 tonne CO2/day capture plant (DOE project DE-FE0031587). The Large Pilot represents an approximate 10-fold scale-up from MTR's previous small pilot membrane system field tests. We are currently in the third and final phase of this project. In Phase I, which began in 2019, the overall feasibility of the capture process was determined, and the host site was identified. In Phase II, a Front End Engineering and Design (FEED) study was conducted, a schedule for system construction and operation was prepared, and permits were obtained.

The Large Pilot is being installed and will be operated at the Wyoming Integrated Test Center (WITC) in Gillette, Wyoming. The WITC is a new, dedicated post-combustion carbon capture test facility attached to Basin Electric's Dry Fork Station (DFS); a single unit, coal fired power plant.

This presentation will describe all Large Pilot process elements that are part of MTR's full-scale commercial process design such as the flue gas pretreatment and conditioning; two stages of membrane unit operations; and a CO2 purification unit. Just as important, the Large Pilot will also feature MTR's containerized membrane skids and planar membrane modules, both in their final commercial form factor.

Construction work began in May 2023 with a groundbreaking ceremony and will continue through summer 2024. When operational in 2024, the Large Pilot will be the largest membrane-based capture plant in the world. The results from Large Pilot testing will be used to refine the MTR CO2 capture process for a potential full-scale demonstration plant at DFS as well as other industrial sources such as cement plants and steel plants.

8:50 AM
Oral 152 - Designing membranes using bottlebrush poly(1,3-dioxolane) acetate for CO2 capture

Fathy Attia (The State University of New York at Buffalo)*; Thien N Tran (University at Buffalo, The State University of New York); Erda Deng (State University of New York at Buffalo); Bhanuprakash Valluri (University at Buffalo); Liang Huang (The State University of New York at Buffalo); Haiqing Lin (The State University of New York at Buffalo)

Poly(ethylene oxide) (PEO) based polymers are leading membrane materials for CO2/N2 separation because the ether oxygen groups show affinity towards CO2. Herein, we synthesize polymers based on 1,3-dioxolane (DXL) with higher ether oxygen content than PEO, thus improving CO2/N2 separation performance. Specifically, macromonomers of poly(1,3-dioxolane) acrylate (DXLAn) were synthesized through Cationic Ring-Opening Polymerization (CROP) of DXL and polymers (pDXLAn) were synthesized by free radical polymerization. The macromonomers' chemical structure was confirmed via 1H-NMR, and the polymers' amorphous nature was validated using DSC. Pure-gas CO2/N2 separation properties were determined as a function of feed pressure, composition, and temperature, and the effect of the pDXLAn chain-end group on the gas separation performance was also investigated. The transformation of the -OH chain-end group within pDXLAn into acetate (-OCOCH3), as in pDXLAc, disrupts hydrogen bonding, leading to improved polymer chains' flexibility and enhanced gas permeability. For example, converting pDXLA12 to pDXLAc12 increased CO2 permeability from 251 to 337 Barrer, albeit with a slight decrease in CO2/N2 selectivity from 61 to 51. Despite the impressive performance of pDXLAn and pDXLAcn, their free-standing films (300 μm) proved insoluble in typical solvents, hindering the preparation of coating solutions for thin-film composite (TFC) membranes. To overcome this, a soluble high-molecular weight pDXLAc was synthesized via RAFT polymerization. Subsequently, these polymers were successfully fabricated into defect-free TFC membranes, resulting in a CO2 permeance of 1831 GPU and CO2/N2 selectivity of 38 at 22 ºC, comparable to leading state-of-the-art membranes for carbon capture. This study highlights the efficiency of end-group acetylation of pDXLA in designing amorphous polymers for superior CO2 capture. Additionally, the mixed-gas separation properties and performance will also be discussed.

9:10 AM

Oral 153 - Random Poly(1,3-dioxolane)-based Terpolymers for CO2 Separations

Justin J Rosenthal (The University of Texas at Austin)*; Isaac Tan (The University of Texas at Austin); Julian Richardson (University of Texas, Austin); Nathaniel Lynd (The University of Texas at Austin); Benny Freeman (The University of Texas at Austin)

Global climate change as a result of anthropogenic carbon dioxide (CO2) emissions is one of the most pressing issues facing humanity. Massive efforts are being put forth to develop technologies which preferentially separate CO2 from other gases, such as nitrogen (N2) and methane (CH4), for sequestration or other end uses. Membrane based separation of CO2 is widely studied since it is considered to be more energy efficient than conventional thermal separation processes. For industrial viability, membranes must have high CO2 permeability and permselectivity. Polyether membranes based on poly(ethylene oxide) (PEO) have shown desirable CO2/N2 separation performance due to their polar ether oxygen groups which exhibit a strong affinity toward CO2. Increasing the ether oxygen content of polyethers has been theorized to increase the CO2/N2 solubility-selectivity of these polymers. Since poly(1,3-dioxolane) (PDXL) has a higher ether oxygen content than PEO, it is expected to have enhanced separation performance. However, pure PDXL displays high crystallinity which dramatically reduces the gas permeability. Here, random PDXL-based terpolymers were synthesized through copolymerization with epoxides to promote amorphous, rather than crystalline, membranes. A variety of epoxides were employed to yield a range of polymers with different chemical and physical characteristics. Resulting membranes were evaluated for their pure- and mixed-gas transport properties. The developed polymers display favorable CO2/N2 separation properties that are competitive with Robeson's 2008 upper bound. For example, the terpolymer based on propylene oxide displays a CO2 permeability of 330 Barrer and CO2/N2 selectivity of 52 at 35 °C. This study demonstrates that random PDXL-based terpolymers yield amorphous membranes that are highly selective for CO2 and display overall favorable gas separation capabilities.

9:30 AM

Oral 154 - Highly Permeable Rubbery Thin Film Composite Membranes for CO2 Capture from Steel Mills

Lingxiang Zhu (National Energy Technology Laboratory)*; Thien Tran (National Energy Technology Laboratory); Fangming Xiang (NETL); Victor A Kusuma (NETL); Cheick Dosso (Carnegie Mellon University); Hector A Pedrozo (Carnegie Mellon University); Grigorios Panagakos (Carnegie Mellon University); Neil Pergar (U. S. Steel); Brenda Petrilena (U. S. Steel); David Hopkinson (NETL)

The U.S. government is targeting a net-zero carbon-emission economy by 2050, offering an exciting opportunity for membrane-based CO2 capture from various industrial point sources. Given that industrial flue gas has low CO2 partial pressures and high volumetric flow rates, high-permeance membranes are needed to make membrane technology economically viable for large-scale deployment. Thin film composite (TFC) membranes are necessary for this implementation because they can provide high permeance by forming a thin selective layer on top of a porous support. This presentation will report the rational design and fabrication of NETL's highly permeable non-aging TFCs achieved by: synthesizing a high-performance rubbery selective material; developing a high-porosity membrane support; optimizing coating methods to assemble the two materials into scalable membranes; and scaling up membrane supports and TFCs via a roll-to-roll process. The novel rubbery selective material shows mixed-gas CO_2 permeability of 930 Barrer and CO_2/N_2 selectivity of 44, exceeding the 2008 Robeson upper bound. The resulting TFCs yield remarkably high CO_2 permeance of 4,500 GPU and CO_2/N_2 selectivity of 34 at 23 ºC. Moreover, the TFCs exhibit not only excellent performance stability (or non-aging behavior) for 1,000 hours in the lab, but also maintain their separation properties in a 700-hour field test at the U.S. DOE's National Carbon Capture Center using real humid flue gas. This talk will also cover the design, computational fluid dynamic simulation, 3D printing, and permeation testing of plate-and-frame membrane modules for an upcoming field demonstration at U. S. Steel's Edgar Thomson Plant in Braddock, PA. The field test will demonstrate membrane-based CO_2 capture from a blast furnace for decarbonization of steel manufacturing for the first time in the U.S. A mobile gas permeation test unit is currently under design and construction for the field test, which is scheduled to take place in 2025.

9:50 AM

Oral 155 - Investigating PIM-1 and cPIM-1 applications for CCUS under real-world conditions

Sean R McIntyre (surface measurement systems)*; Andrew Foster (University of Manchester); Peter M. Budd (University of Manchester); Paul Iacomi (Surface Measurement Systems); Daryl Williams (Imperial College London)

Carbon capture, utilization, and storage (CCUS) is a costly endeavor. Traditional sorbents such as amines are often energy intensive in their regeneration steps, leading to large operating expenses, and can produce toxic by-products. Using membranes over these legacy technologies would have the advantage of ease of implementation, with lower energy and heat requirements. [1,2] An attractive type of membrane material consists of polymers with intrinsic microporosity (PIMs) – forming flexible and easy to manufacture single component membranes with high gas permeabilities and selectivity for CO_2 over N_2 and O_2 – primary components of flue gas. [3] However, whereas the selectivity and permeability of membranes is often measured in ideal mixture conditions, real flue gas contains contaminants such as water which can impact the transport properties of a membrane. Water can cluster in the "Langmuir" voids causing an obstruction of the pore, decreasing permeability. [3,4] In this work, a novel membrane analyzer was developed to observe the multi-component permeation of flue gas constituents through membranes of PIM-1 and its more hydrophilic carboxylate functionalized version, cPIM-1, to reveal the impact of contaminants and process conditions on the membrane separation efficiency. Humidity was shown to impact both CO_2 permeability and selectivity, decreasing by up to 20% from around 3300 barrer. A decrease in water diffusion with increasing relative humidity has shown a build-up of water within the membrane channels. The work also revealed how modifying the PIM-1 structure through carboxylation can change the influence of humidity on the separation of CO_2 from flue gas constituents such as N_2 and H_2O. Further temperature studies provided permeation and diffusion activation constants, matching well with literature values at -7.24 and +6.93 KJ/mol respectively. [5] This study highlights the critical influence of gas-phase contaminants on membrane permeation and selectivity.

10:10 AM

Oral 156 - Retarded O_2 transport in Co^{2+}-coordinated supramolecular polymer networks for membrane CO_2/O_2 separations

Haiqing Lin*

CO_2/O_2 separations are of great interest for carbon capture from fossil fuel-derived flue gas, which contains 4-16% O_2 that needs to be separated from CO_2. Dissociated Co^{2+} ions in liquids and polymers have been demonstrated to reversibly react with O_2 and increase O_2 permeability. However, we find a series of Co^{2+}-coordinated supramolecular polymer networks (SPNs) with enormous O_2 sorption but retarded diffusion, leading to superior CO_2/O_2 separation properties. Specifically, $Co(BF_4)_2$ is dissolved by a cross-linked poly(ethylene oxide) (XLPEO), as validated by Fourier transform infrared (FTIR) spectroscopy, X-ray diffraction (XRD), and differential scanning calorimetry (DSC). The dissociated $Co2+$ ions increase O2/CO2 solubility selectivity but decrease its diffusivity selectivity. For example, adding 6.4 mass% Co(BF4)2 in XLPEO increases O2 solubility by 35 times and O2/CO2 solubility selectivity from 0.12 to 4.8, but it decreases O2/CO2 diffusivity selectivity from 0.43 to 0.0056, leading to CO2/O2 permeability selectivity of 38, above Robeson's upper bound and superior to state-of-the-art polymers. This study unravels an exciting platform of metal ions-coordinated supramolecular networks for various molecular separations by harnessing strong affinity but retarded diffusion.

10:30 AM

Oral 157 - A Membrane Contactor Enabling Energy-Efficient CO2 Capture from Point Sources with Physical Solvent

Nicholas Gregorich (Oak Ridge National Laboratory)*; Mary Irwin (Oak Ridge National Laboratory); Zachary Coin (Oak Ridge National Laboratory); Stephen JA DeWitt (Arkema Inc); Ramesh Bhave (Ornl); David Sholl (Oak Ridge National Laboratory); Syed Islam (Oak Ridge National Laboratory)

Greenhouse gas contributions to global climate change have generated interest in separating carbon dioxide (CO2) from wet flue gas streams. To address this challenge, absorption of CO2 via environmentally friendly, green solvents, has grown as a useful technique. Solvent-based capture systems have emerged in the carbon capture space due to their reusability, high absorption capacities, and favorable economics. Compared to commercially available carbon capture processes that utilize ionic liquid and amine-based technologies, membrane-based carbon capture with diethyl sebacate (DS) is cheaper, energy efficient, and presents high capacity for industrial level carbon capture use. The use of DS as a solvent for pre- and post-combustion carbon capture has many advantages over other solvents such as high hydrophobicity, low viscosity, low vapor pressure, high CO2 solubility, high CO2 selectivity, and being more commercially available in larger quantities as compared to deep eutectic solvent, a common green solvent. Despite its advantageous properties, little research has investigated DS as a solvent for carbon capture. To absorb CO2 from flue gas streams, a scalable, energy-efficient, hollow fiber membrane (microporous polypropylene and polyvinylidene fluoride) contactor (HFMC)-based process was designed with low-cost and high surface area to provide high interfacial area for effective CO2 capture. Single gas measurements in this system demonstrated effective capture of CO2 while rejecting nitrogen. A purity of 95.3% CO2 was achieved using DS, with a permeate flux over one magnitude greater than deep eutectic solvent in this same system. Results from this work underscore the importance of utilizing green solvents in HFMC-based separation processes for effective carbon capture and provides a pathway toward practical deployment.

Specific Ion Separations

8:30 AM

Oral 158 - Elucidating selectivity mechanisms of anionic pollutants in nanofiltration membranes

Idit Shefer (Technion)*; Razi Epsztein (Technion)

Nanofiltration (NF) membranes show great potential to address future water challenges. To achieve maximum benefit from NF, a fundamental understanding of the transport mechanisms in NF membranes is required. In this study, we explored the selectivity trends of common anionic pollutants (i.e., nitrate and perchlorate) compared to chloride and examined their transport mechanisms in polyamide NF membranes. Permeation experiments show that nitrate and perchlorate, despite being polyatomic and larger, permeate the membrane faster than the monoatomic, spherical chloride, suggesting that other mechanisms beyond size and charge exclusion govern the separation. Applying Eyring's transition-state theory to elucidate molecular phenomena that induce such selectivity revealed ion-specific mechanisms as ion dehydration. To further explore the influence of dehydration, we introduced various organic aliphatic alcohols into the feed solution. For chaotropic molecules like perchlorate and nitrate, the inclusion of aliphatic alcohols intensifies their chaotropic characteristics, potentially augmenting their capacity to dehydrate and exhibit enhanced permeation through the membrane. In contrast, the permeability of chloride, which is more kosmotropic compared to perchlorate and nitrate, remains relatively constant upon alcohol addition. The difference between the behaviors of these two types of anions becomes more apparent through the reduced activation energy of perchlorate and nitrate when alcohols are added, while the activation energy of chloride is of similar values in both water and alcohol mixture solutions. Notably, we have witnessed that this effect depends on the presence of the other ions, presumably due to their counter effect on water structuring. Our findings illuminate the significance of ion dehydration in membrane permeability and promote the understanding of required structural and chemical criteria towards membranes with greater selectivity to target pollutants.

8:50 AM

Oral 159 - A study of NF membrane assessment as pretreatment for lithium recovery from salt-lake brine

Juan Zhai (Texas Tech University)*; Yuexiao Shen (Texas Tech University); Mahdi Malmali (Texas Tech University); Rajesh Khare (Texas Tech University); Amrika Deonarine (Texas Tech University); Shubhra Bhattacharjee (Texas Tech University); Adegbola Balogun (Texas Tech University)

As global demand for lithium grows rapidly, lithium extraction has been gradually shifted to lithium - enriched brines.However, lithium coexists with other abundant minerals in the brine, including Mg2+, Ca2+, Na+, and K+. The challenge of extracting lithium from the complex brine has attracted a lot of interests in advanced strategies such as nanofiltration (NF). In this project, we focused on potential new technologies for extracting lithium from one of the promising lithium resources in the United States, Salton Sea geothermal brine. We used simulated Salton Sea geothermal brine as feed solution to investigate the mono-/multivalent selectivity and water flux of widely used commercial NF membranes (DL, NF270, and NF90). We found that low flux, poor monovalent vs. divalent cation selectivity (M+/M2+) of these membranes under high salinity brine conditions, while dilute brine (20~40 g/L) solution can improve lithium separate efficiency. pH was barely influential membrane performance (i.e., water flux, rejection, and selectivity) under high salinity feed solution. Molecular dynamics (MD) simulation confirmed that the existence of counter ions in the polymer membranes under high salinity conditions increased the water-ion interaction (e.g., hydrogen bonding), thus reducing the water mobility; while the salt mobility deterioration was found to be relatively insignificant. Considering the high divalent cation concentration (i.e., Ca2+ up to 25 g/L), it is necessary to employ a combination of NF and crystallization. A preliminary cost analysis has been conducted. We proposed to place a crystallizer in front of NF to remove most of the divalent cations. Then, the treated brine needed to be diluted to a range of 20~40 g/L and 3-stages NF can achieve the final molar Li/Mg and Li/Ca ratio both > 99.

9:10 AM

Oral 160 - Understanding the dopant effects on the selective transport of cations in conductive polymers

Meng Shen (California State University, Fullerton)*

Polypyrrole (PPy), a conductive polymer, has been used to modify exchange membranes (IEM) to improve the selectivity of anions. However, the atomic mechanisms and the limit of the enhanced selectivity of cations by PPy-coating is still unknown. We hypothesize that the controlled improvement of ion selectivity can be achieved by manipulating the polypyrrole charge and structure by dopants. We use molecular dynamics to investigate the effect of dopants on the selective transport of cations, for example, Li+ and Ca2+. We elucidate how dopants influence the polypyrrole morphology and the partition and diffusion of cations in the membrane by analyzing the polymer morphology, ion hydration, interaction energy landscape and activation barrier for transport. The research sheds light on the mechanisms and the limit of cation selectivity improvements by tuning the polymer electrostatics and structures by dopants.

9:30 AM

Oral 161 - Ion-selective Membranes for Critical Materials Separation

Ngoc Bui (The University of Oklahoma)*

This study aims at the design and fabrication of an emerging adsorptive membrane platform for transition metal ion capture. Specifically, the membrane composes of a selective layer built from stacks of zinc imidazole salicylaldoxime (ZIOS) crystalline nanosheets on a polyvinylidene fluoride (PVDF) supporting layer pretreated with polydopamine and polyethyleneimine. The dimensional and morphological textures of ZIOS layers were tuned by varying the concentrations of poly(vinyl alcohol) (PVA, Mw = 146,000-186,000 g/mol) in the reaction media. Additionally, textural properties (e.g., morphology, crystal size, particle orientation, and uniformity) and the adhesion of the ZIOS selective layer to the PVDF-PDA-PEI supporting substrate were controlled by tuning the synthetic conditions. Our results show that these ZIOS/PVDF-PDA-PEI membranes feature distinct ion transport and adsorption efficiency, which are critical for selective metal ion capture. The Cu2+/Ni2+ separating factor was calculated to be 193. We use in-situ Raman spectroscopy to elucidate the underlying mechanisms governing the preferential metal ion uptake by the adsorptive membranes, of which the ZIOS layer functions as a selective layer in discriminating the guest metal cations. Coupled with first-principles calculations, results from this study synergistically provide insight into mechanistic understanding of metal ion separating behaviors of an emerging membrane platform based on ZIOS, a crystalline adsorbent system that we designed specifically for copper/nickel separation.

9:50 AM

Oral 162 - Aligned macrocycle pores in ultrathin films for accurate molecular sieving

Zhiwei Jiang (Queen Mary University of London)*; Ruijiao Dong (Shanghai Jiao Tong University); Austin Evans (University of Florida); Niklas Biere (Bielefeld University); Mahmood Abdulsalam (Imperial College London); siyao li (Imperial College London); Dario Anselmetti (Bielefeld University); William Dichtel (Northwestern University); Andrew Livingston (Queen Mary University of London)

Polymer membranes are widely used in separation processes including desalination, organic solvent nanofiltration and crude oil fractionation. Nevertheless, direct evidence of subnanometre pores and a feasible method of manipulating their size is still challenging because of the molecular fluctuations of poorly defined voids in polymers. Macrocycles with intrinsic cavities could potentially tackle this challenge. However, unfunctionalized macrocycles with indistinguishable reactivities tend towards disordered packing in films hundreds of nanometers thick, hindering cavity interconnection and formation of through-pores. Here, we synthesized selectively functionalized macrocycles with differentiated reactivities that preferentially aligned to create well-defined pores across an ultrathin nanofilm [1]. The ordered structure was enhanced by reducing the nanofilm thickness down to several nanometers. This orientated architecture enabled direct visualization of subnanometre macrocycle pores in the nanofilm surfaces, with the size tailored to ångström precision by varying the macrocycle identity. Aligned macrocycle membranes provided twice the methanol permeance and higher selectivity compared to disordered counterparts. Used in high-value separations, exemplified here by enriching cannabidiol oil, they achieved one order of magnitude faster ethanol transport and threefold higher enrichment than commercial state-of-the-art membranes. This approach offers a feasible strategy for creating subnanometre channels in polymer membranes, and demonstrates their potential for accurate molecular separations.

Reference:

1. Jiang, Z., Dong, R., Evans, A.M., Biere, N., Ebrahim, M.A., Li, S., Anselmetti, D., Dichtel W.R., Livingston, A.G., Aligned macrocycle pores in ultrathin films for accurate molecular sieving. Nature 609, 58–64 (2022)."

10:10 AM

Oral 163 - Tailoring Interlayer Spacing and Surface Chemistry of MXene Free-standing Membranes for Selective Separation of Precious Ions from Water

Ahmed Al Mayyahi (the University of Alabama)*; Milad Rabbani Esfahani (The University of Alabama)

Titanium carbide MXene free-standing membranes (MFMs) have garnered significant attention for their efficiency in ion separation from water, owing to their aligned nanochannels and unique surface chemistry. However, their application in the selective separation of precious ions is seldom and remains to be experimentally explored. In this study, we focused on adjusting the interlayer spacing of MFMs and, consequently, the size of their nanochannels to either match or exceed the diameter of the desired ions. This adjustment facilitates the permeation of precious ions, including Y2+, Cr2+, and Ce2+, through the membrane while rejecting other ions, allowing for their selective separation. We investigated the correlation between membrane surface chemistry and ion selectivity, showcasing the potential of controlling MXene surface chemistry to enhance the efficiency of selective ion separation. Additionally, we extended the lifetime and stability of MFMs through in situ cross-linking via thermal treatment. Our findings pave the way for the practical implementation of these membranes in the selective separation of precious ions, a crucial process for technologies essential to fostering a circular economy and achieving a clean energy future.

10:30 AM

Oral 164 - Selective Ion Conducting Polymers for Non-Aqueous Redox Flow Battery Applications

Patrick McCormack (UVA), Gary Koenig (UVA), Geoffrey Geise (UVA)*

Providing sustainable supplies of clean energy is a critical global challenge for the future. As intermittent renewable energy sources are increasingly deployed, grid-scale energy storage solutions are increasingly needed. One approach to addressing this challenge is to use flow battery technology to store and deliver grid-scale amounts of energy. Non-aqueous redox flow batteries can be operated at higher voltages and energy densities compared to aqueous systems and, as such, may offer high volumetric energy density compared to other flow batteries. A significant challenge facing non-aqueous flow batteries, however, is the lack of selective membrane separators engineered for the unique challenges of non-aqueous electrochemical systems. Suitable membranes for non-aqueous redox flow battery applications must be stable in aggressive solvent environments, offer high conductivity, and provide selectivity to prevent cross-over of redox active molecules. Here, we report the synthesis and characterization of a series of negatively charged ion conductive polymeric membranes for non-aqueous flow battery applications. Fixed negative charges were added to a poly(phenylene oxide) backbone via a custom sulfonate group-based side chain. Lithium ion conductivity and ferrocene permeability properties were characterized to evaluate the selectivity of the membrane for ion transport relative to redox active molecule cross-over (as ferrocene is a representative redox active molecule). The polymers exhibited combinations of conductivity and selectivity that are favorable compared to other Nafion-based materials, and the materials appear to be dimensionally stable in a non-aqueous electrolyte over a period of several months. Redox active molecule cross-over was analyzed within a thermodynamic regular solution framework to highlight how thermodynamic factors contribute to cross-over properties. Ultimately, the data suggest a decoupling of ion and redox active molecule transport that could inform future efforts to develop advanced non-aqueous redox flow battery membrane separators. This presentation discusses the transport properties and stability of these materials that show promise for non-aqueous flow battery applications.

Scaling Up: Bridging the Gap Between Lab and Industry

8:30 AM

Oral 165 - Electrosprayed membranes for CO2-selective gas separations

Noah Ferguson (University of Connecticut)*; Arjun DeJesus (University of Connecticut); Srivatsa Bettahalli (University of Connecticut); Jeffrey R McCutcheon (University of Connecticut)

Industry increasingly turns to membranes for the purification of natural gas, upgrading of biogas, and postcombustion capture of carbon dioxide. This separation technology boasts the attractions of modularity, compact footprint, and reduced mechanization. Yet dramatic advances in membrane materials science have outpaced those of industrial membrane fabrication, which has changed little in the last forty years. Here we present electrospray deposition, a next-generation fabrication technology, for producing thin film composite (TFC) and mixed matrix gas separation membranes. Electrospray employs an electric field between the spray nozzle and a deposition surface. Charged droplets of polymer dope experience Coulombic fission during flight, becoming a fine spray able to yield membranes only tens of nanometers thick. Such membranes can be made from many solution-processable polymers, scalably and repeatably. Our present research employs the poly(ether block amide) copolymer Pebax-1657 as the thin CO2-selective layer, and poly(dimethylsiloxane) as a permeable gutter layer. Additionally, we investigate electrosprayed mixed matrix membranes (MMMs) as a means of boosting permeance and selectivity by harnessing the high internal volume and size-exclusive apertures of the zeolitic imidazolate framework ZIF-8. These MMMs, produced by co-spraying Pebax and ZIF-8, are intended to mitigate particle agglomeration by exploiting the inherent dispersive effects of electrospray. High permeance TFC and MMMs made with electrospray technology may make industrial CO2 separations cheaper and simpler, and help establish membranes as a leading technology for worldwide decarbonization.

8:50 AM

Oral 166 - Simplified Synthesis Methods for Membrane-Protein Nanosheets and Biomimetic Membranes

Ronald J Vogler (The University of Texas at Austin)*; Katherine Kimball (The University of Texas at Austin); Dominic Bujanos (The University of Texas at Austin); Raman Dhiman (The University of Texas at Austin); Benny Freeman (The University of Texas at Austin); Manish Kumar (University of Texas at Austin)

Membrane proteins and artificial channels are promising membrane materials due to their high throughput and precise selectivity. Incorporating these channels and amphiphilic block-copolymers into 2D nanosheets and depositing them onto porous membranes has yielded composite nanofiltration membranes with water permeances 1-2 orders of magnitude higher than commercial membranes of similar selectivity. However, synthesizing these 2D nanosheets and membranes is time-consuming and complex. In this work, simplified synthesis methods are investigated. 2D nanosheets containing Outer Membrane Protein F (OmpF) and poly(ethylene glycol-b-caprolactone) (PEG-PCl) were synthesized by mixing detergent-solubilized OmpF and PEG-PCl, fluctuating the temperature between 25°C and 60°C for 12.5 hours, and diluting the detergent (Octyl-POE) below its critical micelle concentration. TEM indicated the formation of 2D materials with a hexagonal crystal structure. Compared to materials prepared using the conventional detergent-dialysis method, these OmpF/PEG-PCl nanosheets were synthesized ~10x faster and were 1-2 orders of magnitude larger, having areas up to 390 μm^2. We hypothesize that the rapid formation of these large OmpF/PEG-PCl nanosheets was enabled by the melting of PEG-PCl and the temperature-driven phase separation of the detergent. A simplified approach for synthesizing flat-sheet membranes from OmpF nanosheets was also investigated. Instead of performing a 6-12 step layer-by-layer deposition of OmpF nanosheets and poly(ethyleneimine) (PEI), OmpF nanosheets and PEI were mixed and convectively deposited onto an ultrafiltration support membrane in one step. The resulting membrane had a water permeance of 45 LMH/bar and an MWCO of ~650 Da, which is in good agreement with results corresponding to the layer-by-layer method. Together, these simplified synthesis methods for 2D nanosheets and membranes represent an advancement in the scalability of nanosheet-based biomimetic membranes.

9:10 AM

Oral 167 - Evaluation of spacer-induced hydrodynamic mixing using particle image velocimetry: Impact on membrane distillation performance

Hadis Matinpour (Texas Tech University); Mahdi Malmali (Texas Tech University)*; Bosong Lin (Texas Tech University)

Net-type spacers are used to induce hydrodynamic mixing within the feed channel of crossflow membrane filtration, aiming to alleviate TP and CP effects on the membrane surface. This, in turn, enhances heat and mass transfer across the membrane. In this presentation I will discuss on examining the influence of spacer geometry on fluid flow and heat/mass transfer within the feed channel. The particle image velocimetry (PIV) technique is employed to measure flow velocity in the feed channel equipped with net-type spacers. Flow characteristics are described across a wide range of Reynolds numbers in terms of "time-averaged mean velocity" and "turbulence kinetic energy" (TKE). The PIV results have revealed distinct profiles of mean velocity and TKE at various channel heights with bi-planar spacers. Notably, the spacer with the largest hydrodynamic angle of 120° (Net120) proves to be the most effective in promoting flow unsteadiness, as validated by the highest channel-averaged TKE among the selected spacers, which corresponds to the greatest pressure drop across the feed channel.

Furthermore, the impact of spacer-induced mixing on heat/mass transfer across the membrane was assessed through flux measurements in vacuum membrane distillation (VMD) experiments. Compared to other spacers, Net120 demonstrated a relatively higher flux due to improved hydrodynamic mixing. These results affirm the effectiveness of the experimental approach in elucidating the influence of key spacer design parameters on hydrodynamic mixing. To better predict the convective heat transfer in the spacer-filled channel, we proposed a modified Nusselt number correlation in the spacer-filled channel in VMD. Our study illustrates a useful method to evaluate the impact of spacer geometry on hydrodynamic mixing, which can be used to identify key spacer design parameters.

9:30 AM

Oral 168 - A Fractal Analysis Based Metrology to Quantify Porous Structure of Membranes

Riccardo Lovison (University of Colorado Boulder)*; Yifu Ding (Membrane Science, Engineering, and Technology (MAST) Center, Department of Mechanical Engineering, University of Colorado Boulder); Christina Carbrello (MilliporeSigma)

Accurate characterization of intricate 3D porous membrane structures is critical for the development of advanced membranes. Currently, pore structures (porosity, pore size and tortuosity) are routinely quantified with techniques based on mass and permeation (gas, liquid, or other types of challenging solutions) measurements, as well as microscopic imagine techniques. Precisely defining these parameters in asymmetric porous membranes represents a challenge due to limitations in resolution and reliability for in-depth measurements. This study explores a new approach of using femtosecond laser ablation to obtain high-quality In-Plane Cross-Section (IPCS) images at submicron intervals. Polyether sulfone (PES) membranes with asymmetric structure and 200 nm nominal pore size have been tested. Polypropylene has been used as pore-preserving agent to minimize thermal damage during the sectioning procedure. With optimized laser ablation parameters (laser fluence and number of passes), IPCS images of the PES membranes at varying depths were successfully obtained.

Next, the study employs Fractal Analysis (FA) to quantify the membrane's morphology, offering new insights into mass distribution at different length scales. FA has demonstrated its effectiveness in characterizing random porous media within hydrology, soil science, and even oil recovery domains. However, its application remains unexplored in the field of membrane science, specifically in delineating membrane properties such as permeability and selectivity. To this end, the adaptive thresholding technique is employed for image binarization, followed by the utilization of the box counting method to measure pore area and mass fractal dimension. The study reveals variations in fractal dimensions at different depths within the asymmetric PES membrane. The introduction of the FA holds the potential to enhance quantitative analysis of membrane structures, beyond the commonly used porosity, pore size and distribution.

9:50 AM

Oral 169 - Design of Experiments to Optimize the Polyimide Membrane Structure

Margaret Lee (Pall Corporation)*; Siavash Darvishmanesh (Pall corporation)

Polyimide (PI) is a high-performance polymer whose excellent thermal and chemical stability have made it a favorable candidate for use in membrane separation. Application studies for PI have primarily been for dense membranes in gas separation, but there is interest in developing porous membranes to utilize PI's properties in liquid filtration as well. This abstract highlights Pall Corporation's ongoing work into the development of porous PI membranes for use in various filtration applications. A custom designed experiment with six factors was used to optimize the structure of a porous PI membrane prepared using nonsolvent induced phase inversion (NIPS). The structure of the PI membrane was characterized using scanning electron microscopy (SEM) to evaluate the cross-sectional morphology, and Porolux to measure the average pore size and solvent flow. Understanding the effects of the design factors allows for improved control over membrane structure, and within the defined design space, symmetric and asymmetric porous PI membranes with a mean pore size of 20-40 nm and high solvent flow were fabricated. In combination with its thermal and chemical stability, this makes porous PI membranes a suitable candidate for a range of filtration applications including microelectronics and the chemical industry.

10:10 AM

Oral 170 - Spiral Wound Module Development and Scale-up for Olefin-Paraffin Separations with OptipermTM Membrane System

Jacob Wargo (Compact Membrane Systems)*; Christine Parrish (Compact Membrane Systems); Brandon J Burghardt (Compact Membrane Systems); Bill Charlton (Compact Membrane Systems)

Compact Membrane Systems has developed the novel Optiperm™ membrane platform for decarbonizing today's industry across two applications: olefin paraffin separation and carbon capture. The separation of light olefins from paraffins is one of the most energy intensive in the world. The state of the art separation method, distillation, remains unchanged over decades. Membranes promise a more modular, energy efficient and capital efficient method for separation. However, they have historically been hampered by low performance and stability. CMS has developed a transformational membrane technology that is being demonstrated at commercial scale.

Crossing the chasm between lab results and successful field trials, is the largest challenge for membranes in displacing incumbent technology. In this talk, we will discuss the process and challenges of spiral wound module scale-up, manufacturing and operation. We will also explore design considerations gleaned from commercial scale pilot operations, and how well refined lab scale testing and development can be crucial in designing a system that can meet a customers desired specifications for a commercial application. The sensitivity of selectivity and permeance to field operating conditions will be detailed, along with the operation consideration to ensure reliable product quality.

The first Optiperm™ membrane utilizes a silver embedded polymeric material to selectively transport olefins. This material has demonstrated over 3 years of longevity in laboratory operation and has shown high selectivity and permeance. Additionally, Optiperm™ membranes were first trialed at the Delaware City Refining Company in over 300 days of field operations validating their robustness in a plant environment. CMS will also give an update on the commercial demonstration of the technology with industrial partner Braskem. This is the first commercial scale operation of Optiperm™ membrane modules under field conditions using actual process streams.

10:30 AM

Oral 171 - Effect of polymer molecular weight on the long-term process stability of crosslinked polybenzimidazole (PBI) membranes

Adam Oxley (Exactmer Ltd)*; Andrew Livingston (Queen Mary University of London)

Polymer molecular weight has a significant impact on non-solvent induced phase inversion membranes. It is well established that the viscosity of the dope solution is dependent on the polymer molecular weight, which in turn affects the rate of phase inversion. However, the long-term stability of the resultant membranes made from different polymer molecular weights has not been extensively studied.

In this work, PBI membranes were manufactured from two different polymer molecular weights: the standard molecular weight of PBI polymer, used almost universally across literature for organic solvent stable membranes (~27,000 g mol-1), and an untested ~ two-fold higher molecular weight (~60,000 g mol-1). After crosslinking, all PBI membranes were chemically and mechanically stable. However, during continuous long-term pressurised filtration in DMF, the standard molecular weight PBI membranes suffered slow permeance decline over time. This was attributed to rearrangement of the crosslinked polymer chains. Increasing the extent of crosslinking reduced the rate of membrane permeance decline but did not provide a fully process stable membrane. Using a higher molecular weight PBI, and a similar extent of crosslinking, resulted in a membrane with constant permeance and excellent process stability, even during continuous DMF filtration at 120 °C. This was attributed to the increased interchain interactions and entanglement in the high molecular weight polymer, which reduced the rate of chain rearrangement and compaction. This work demonstrates the importance of polymer molecular weight for reducing membrane ageing and providing process stability in organic solvents.

Contaminants Removal from Water I

8:30 AM

Oral 172 - Structure-property relationships in intercalated graphene oxide nanofiltration membranes for enhanced separation of salts and organics from aqueous streams

Muskan Sonker (Georgia Institute of Technology)*; Sankar Nair (Georgia Institute of Technology); Kaung Su Khin Zaw (Georgia Institute of Technology); Hazel P. Dhruve (, Georgia Institute of Technology); Mahsa Abbaszadeh (Albemarle); Madeline P. Garell (Georgia Institute of Technology); Michael B. Salerno (Phillips 66 Research Center); Marta Hatzell (Georgia Institute of Technology); Meisha L. Shofner (Georgia Institute of Technology)

Graphene oxide (GO) membranes are emerging as a viable new technology for separation of complex aqueous streams demonstrating better chemical resistance than polymeric membranes. Prior work at Georgia Tech has developed scalable GO membranes that are robust in a variety of harsh operating environments. Additionally, GO membranes intercalated by conjugated polyaromatic molecules showed enhanced rejection of salts as well as excellent rejection of organics (including hydrocarbons). However, the mechanisms underlying these findings are complex. In this talk we discuss key microstructure-separation property relations of GO nanofiltration membranes that are intercalated/functionalized by p-conjugated polyaromatic intercalants. Initially, we used a representative intercalant (7-amino-8-methylphenothiazin-3-ylidene)-dimethylammonium chloride, also known as toluidine blue O (TBO) to determine structure-property relations by a combination of experimental techniques including XRD, UV-Vis, surface zeta potential, fluorescence spectroscopy, and permeation measurements. We extended our learnings to develop GO nanofiltration membranes with enhanced rejection of monovalent salts and hydrocarbons for applications such as recycling of hydrocarbon-containing refinery wastewaters.

8:50 AM

Oral 173 - Removal of Dodecyl Benzenesulfonic Acid by Amine-Modified Membrane Adsorbers

Hadi Rouhi (University of Alabama)*; Nick Vacca (University of Alabama); Jess Moskey (University of Alabama); Steven T Weinman (The University of Alabama)

The increasing worldwide need for clean water along with challenges like scarcity driven by urbanization, industrialization, and climate disruptions, presents a significant dilemma. The presence of dodecylbenzene sulfonic acid (DBSA) from detergents in water sources further compounds environmental concerns, necessitating effective remediation strategies. In this work, polyvinylidene difluoride (PVDF) membranes were modified via a redox-initiated radical polymerization to impart adsorption capabilities to the membranes. We hypothesized that if PVDF membranes were modified with monomers containing tertiary amine groups, then the PVDF membrane could effectively adsorb DBSA and be regenerable with a sodium hydroxide (NaOH) solution. The monomer used in this study was 2-(Dimethylamino)ethyl methacrylate (DMAEMA) and the comonomer was methacrylic acid (MAAc). Different techniques such as ATR-FTIR, SEM, contact angle goniometry, and BET were used to characterize the modified membrane. By introducing tertiary amine groups onto the PVDF surface, effective DBSA adsorption was achieved, resulting in an increased DBSA removal rate from 8.9% for unmodified to 98.9% for the best performing modified PVDF membranes. The adsorption kinetics followed the pseudo-second-order model and the adsorption isotherms correlated well with the Freundlich model. Thermodynamic analysis revealed the adsorption process to be exothermic. Dynamic adsorption tests demonstrated 60-65% removal efficiency (RE) using one single membrane coupon in a dead-end filtration cell. The 5-cycle regeneration experiments showed that RE gradually decreasing from 97% during the first cycle to 51% by the fifth batch adsorption cycle. This study highlights how surface modification techniques have the potential to confer additional separation capabilities to membranes, aiding in the elimination of hazardous surfactants from polluted water sources.

9:10 AM

Oral 174 - Thin Film Nanocomposite Membranes Embedded with ZIF-67 for Desalination and Boron Removal

Suer Kurklu Kocaoglu (The University of Texas at Austin)*; Benny Freeman (The University of Texas at Austin); S. Birgül Tantekin-Ersolmaz (Istanbul Technical University)

Thin film composite (TFC) membranes are the state-of-the-art for reverse osmosis (RO) processes which is the leading desalination technology. Although TFC membranes provide 99% salt removal, low permeability, and removal of small and neutral molecules like boron is an emerging problem. Boron is present in seawater mainly as boric acid and high boron content in the produced water cause health problems for humans. The properties of TFC membranes are enhanced by embedding nanoparticles either in the support membrane or polyamide layer. This improves selectivity as well as water flux and antifouling properties. Among nanomaterials, metal organic frameworks (MOF) like ZIF-8, UiO-66, ZIF-67 have been studied for boron adsorption from aqueous solutions. Previous studies showed that ZIF-67 has the highest boron adsorption capacity and enhances water permeability. Hence, incorporating ZIF-67 into polyamide reverse osmosis membrane could increase permeability and boron selectivity at the same time without compromising salt rejection.

Here, we reveal the influence of incorporating ZIF-67 nanoparticles into TFC RO membranes on their desalination and boron rejection performance. ZIF-67 is synthesized by solvothermal method and characterized by XRD, DLS Zetasizer, TGA and BET analysis. TFN membranes are prepared by interfacial polymerization using m-phenylenediamine (MPD) and trimesoyl chloride (TMC). ZIF-67 is incorporated into the polyamide layer by either dispersion into solution or in-situ synthesis. The membrane is characterized by FESEM, EDX, XPS, and contact angle measurements. Water permeability is conducted in a cross-flow RO test cell at 15.5 bar and 2000 ppm NaCl and 5 ppm boron feed solution is used as feed solution for separation performance. TFN membrane prepared using monomer concentrations of 0.5%w/v TMC, and 2%w/v MPD by dispersing ZIF67 in MPD solution increased boron removal by 24% and maintained 99.1% salt rejection with 9.8 LMH water flux at pH 9.5.

9:30 AM

Oral 175 - Unlocking the power of a gutter layer: A new frontier in dye-desalination using amine-crosslinked nanoporous graphene oxide (aHGO) platform

Ameya Tandel (State University of New York, Buffalo, NY)*; Haiqing Lin (The State University of New York at Buffalo)

We have developed a novel method to modify the surface of commercial polyethersulfone (PES) support membranes using polydopamine (PDA) as an adhesive layer and amine-crosslinked holey graphene oxide (aHGO) as a selective layer for dye removal from textile wastewater. The textile industry needs membranes with high water permeance, low salt rejection, and complete dye rejection. We carefully improved the surface characteristics of PES (pore size: 450 nm) using PDA to enhance the alignment of aHGO nanosheets to achieve defect-free membranes with thin selective layers. We studied the impact of dopamine concentration (g/L), dopamine exposure time (mins), and aHGO selective layer thickness (nm) on the membrane structure and performance. Increasing PDA surface modification time can cover the PES surface completely, improving support surface features and dye rejection rate verified with SEM images and performance data. For instance, increasing PDA exposure time from 0 to 30 mins increases direct red-80 (DR, 1373 g/mol) rejection from 79.1 to 99.6% for 50 nm aHGO membranes. Moreover, increasing the selective layer thickness from 50 to 120 nm (4 times decrease than without PDA exposure) for a fixed PDA exposure (2 g/L, 15 mins) increases DR rejection from 96 to 99.9% with a minor decrease in dye water permeance from 150 to 120 LMH/bar. Our optimized aHGO/PDA coated flatsheet membranes offer excellent dye water permeance of 100 LMH/bar, dye rejection of 99.9%, Na2SO4 rejection of <10 %, and an exceptional dye-salt separation factor of 900, better than many GO membranes reported in the literature. We also determine the relationship between PDA exposure time and PDA concentration on the membrane performance. For example, 2 g/L PDA with 15 mins exposure performs similarly to 7.5 mins of 4 g/L PDA exposure, creating a flexible platform for membrane fabrication. We have also fabricated the aHGO on hollow fiber membranes to demonstrate the commercialization potential.

9:50 AM

Oral 176 - Experimental and Modeling Results for the Rejection of Ultrashort Per-and Polyfluoroalkyl Substances

Aron Griffin (Colorado School of Mines)*; Timothy Strathmann (Colorado School of Mines); Christopher Bellona (Colorado School of Mines)

Per- and polyfluoroalkyl substances (PFAS) are a contaminant class of emerging concern that are ubiquitous in environmental waters. Impending regulations will necessitate installation of advanced treatment technologies for the removal of PFAS at facilities throughout the country. Granular activated carbon and ion exchange resins are effective for PFAS removal from cleaner matrices such as groundwater, but elevated dissolved organic matter, inorganic ions, and other organic co-contaminants present in wastewater inhibits PFAS adsorption, especially short and ultrashort chain PFAS that have been increasingly found at high concentrations in the environment. High pressure nanofiltration (NF) or reverse osmosis (RO) membranes are a potentially more robust technology for PFAS removal, but further work is required to confirm the efficacy of removal of varied PFAS species in wastewater matrices. We evaluated the rejection of PFAS of varied carbon chain length and charge at the bench scale under multiple pH and ionic strength conditions, showing that tight NF membranes are able to reject even the lowest molecular weight C1 compounds. Bench studies evaluating multiple real wastewaters were also performed. Rejection results were fitted with the phenomenological model and the donnan steric partitioning model to predict rejection. Additional experiments utilizing a pilot system operated in a closed circuit configuration to treat lab mixture solutions and real wastewaters found that high PFAS rejection can be achieved in complex matrices and high water recoveries (e.g. PFAS rejection in semiconductor fabrication (FAB) wastewater). These results are important for any practitioner interested in the use of membranes for PFAS treatment.

10:10 AM

Oral 177 - Polymer evolution and fabrication of PVDF membranes for contaminant removal from water

Kaplan, Rachel (University of Kentucky)*; Escobar, Isabel (University of Kentucky)

Membrane technology has proven to be an effective tool for separation processes due to its lower material and fabrication costs, high automation and throughput, and reduced footprint. Phase inversion is the most widely used technique for preparing polymeric membranes where the polymer is dissolved in a suitable solvent and internal or external forces are applied to induce phase separation. The goal of this work is to elucidate the mechanisms within the phase inversion process and the impact of various parameters on the resulting membrane morphology and performance. Polyvinylidene fluoride (PVDF) is the polymer of choice for this study due to its favorable properties such as high mechanical strength, thermal stability, and chemical resistance. This high chemical resistance allows membranes fabricated from PVDF to be regenerated by cleaning with chemical cleaning agents such as chlorine. PVDF membranes were fabricated via non-solvent induced phase separation (NIPS) using N-methyl-2-pyrrolidone (NMP) as the selected solvent and characterized with respect to dope solution viscosity, membrane morphology, surface wettability and composition, permeability, and protein rejection. The results of this study show fabrication method and membrane composition greatly influence the resultant membrane performance. The use of a PVDF co-polymer versus neat PVDF led to more consistent membrane performance. Additives such as polyvinylpyrrolidone (PVP) and polymethyl methacrylate (PMMA) were also explored and led to increased permeability within the membranes. The shear that is induced across the solution surface during casting was found to correlated with membrane performance and morphology. Each of these parameters were varied to optimize membrane performance with respect to bovine serum albumin (BSA) protein and permeability

10:30 AM

Oral 178 - Zr-based MOF-NF-Membrane for Selective Removal of Selenium from Produced water

Sweta A Modak (The University of Alabama)*; Milad Rabbani Esfahani (The University of Alabama)

Oil extraction's produced water, with high salinity (~35000ppm) and elevated selenium (Se) levels (500ppm-300ppt), presents challenges for reuse. Se, with bioaccumulation risks, can be harmful when ingested at levels exceeding 0.05ppm, affecting aquatic life, wildlife, and humans. The use of nanofiltration (NF) membranes has been explored for produced water treatment, demonstrating the effective removal of Se oxyanions at more than 93%. However, NF membranes reject pollutants, demanding an extra step for safe recycling. Adsorptive membranes with Se selectivity adsorb during filtration, removing the need for additional separation. This study introduces MOF-adsorptive membranes, utilizing the advanced properties of metal-organic frameworks (MOFs) to selectively remove selenium (Se) from high salinity and oily waste streams like produced water. Based on the acceptable removal performance of MOF-808 (~75mg/g selenite) and Fe-MOF (~60mg/g selenite), they were selected for grafting on the TFC membrane for the generation of MOF-adsorptive membranes. The mentioned MOFs were grafted on the polyamide layer through chemical grafting. The effect of different operation conditions (MOF dosage, temperature, pH and varying ionic strength) on the physicochemical properties of MOF-NF membrane were studied. The blank TFC membrane exhibited rejection rates of ~85%, ~50%, and ~10% for 2000 ppm Na_2SO_4 and NaCl, and 35000 ppm NaCl. In contrast, the MOF-808-membrane showed ~70%, ~20%, and ~10% rejection for the same concentrations. The water permeability of the blank TFC membrane, Fe-MOF-membrane, and MOF-808-membrane was observed to be ~1.5LMH/bar, ~3.5LMH/bar, and ~4.5LMH/bar respectively. The blank TFC membrane rejected approximately 84% and 88% of 3 ppm selenite and selenate ions, while the MOF-808-membrane rejected around 64% and 83% of the same ions at 3 ppm. The same MOF-membrane adsorbed ~13% and ~8% of 3ppm selenite and selenate ions respectively in 30min at 5LMH/bar permeance.

Abstracts – Oral Presentations, Wednesday Afternoon

Presenting authors are indicated by an asterisk*.

Mixed Matrix and Composite Membranes for Gas Separation

1:00 PM

Oral 179 - UV-Crosslinked Nanocomposite Membranes for Fast and Selective Transport of Small Molecules

Joo-Eon Kim (Korea Research Institute of Chemical Technology (KRICT))*; Mi-Hee Ryu (Korea Research Institute of Chemical Technology (KRICT)); Ahrumi Park (Korea Research Institute of Chemical Technology (KRICT)); Jaesung Park (Korea Research Institute of Chemical Technology (KRICT))

Polymer nanocomposites are emerging as a promising membrane platform for advancing gas separation research. Among various nanofillers, porous organic polymers (POPs) exhibit good compatibility with many organic polymers. This study investigates the impact of hypercrosslinked polymers (HCPs), a specific subtype of POP, on gas transport properties of HCP/polymer nanocomposites. Highly porous HCPs were successfully synthesized and incorporated into a UV-crosslinkable polyimide (PI) matrix. As HCP content increased to 16 wt.%, H2 permeability increased noticeably to 1800 Barrer (six times higher than pure PI), maintaining H2/CH4 selectivity of 20. This improvement overcomes the traditional trade-off limitations in polymer membranes. The characterization of solubility and diffusivity suggests that the incorporation of HCP into the PI matrix may enhance favorable free volume and free volume distribution. Additionally, the nanocomposites underwent post-UV irradiation treatment, resulting in densification and substantial improvement in gas selectivity (e.g., a 3-fold increase in H2/CH4 selectivity from 20 to 60) at the expense of H2 permeability (e.g., 900 Barrer). Notably, in comparison to UV-crosslinked PI exposed to an equivalent UV irradiation dosage, the nanocomposites exhibited a higher gas selectivity. This observation suggests that UV crosslinking of the nanocomposites selectively densifies the matrix to a greater extent. Ultimately, the UV-crosslinked HCP-containing nanocomposites exhibited outstanding gas separation performance, surpassing 2008 upper bounds for H2/CH4 as well as CO2/CH4. The results highlight the synergistic effects of nanofiller additives and crosslinking, improving both the permeability and selectivity of pure PI, establishing its potential as a candidate for H2 separations and CO2 removal. In addition, CO2/CH4 mixed-gas permeation results at elevated pressures also indicate the nanocomposites' good resistance to plasticization.

1:20 PM

Oral 180 - Computational insights from the polymer/MOF interface in Mixed-Matrix Membranes

Moein Hajian Zaker Damghan (Texas Tech University)*; Hyunhee Lee (MIT); Zachary Smith (MIT); Joshua Howe (Texas Tech University)

Gas separation is crucial for a variety of applications, but traditional methods can be energy-intensive and environmentally harmful. Membrane-based separation presents a greener, cost-effective alternative, using selective permeation for efficient gas filtration. However, conventional polymeric membranes often lack the necessary selectivity and permeability for certain applications. Mixed-Matrix Membranes (MMMs), which blend fillers with polymer matrices, have emerged to enhance performance. Metal-Organic Frameworks (MOFs), notable for ability to fine-tune pore environments and interactions at adsorption sites, can improve membrane selectivity without compromising permeability.

Despite advancements in experimental research on MMMs, theoretical developments have lagged due to challenges in accurately modeling these systems. Accurate description of MOFs typically requires electronic detail beyond that in typical polymer models, whereas polymers require system extents in excess of those accessible to typical MOF models. This modeling scale challenge hinders the development of accurate theoretical models of MMMs.

Our study aims to address this by using polymer models limited in extent to investigate electronic effects at MOF-polymer interfaces within MMMs. Focusing on gas diffusion through Zeolitic Imidazolate Framework-8 (ZIF-8) surfaces, interfaced with 6FDA-DAM polymer models, we employ density functional theory and climbing image nudge elastic band calculations. This approach, while not exhaustive, sheds light on electronic properties' influence on interfacial transport barriers affecting separation performance.

Our research offers insights into the structural and electronic properties of MOF/polymer interfaces and explores MMM performance. We seek to enhance understanding of MOF/polymer interfaces, informing the development of materials with improved separation performance and contributing to more efficient, sustainable gas separation solutions.

1:40 PM

Oral 181 - Leveraging anomalous gas transport in thin-film composite membranes based on polymers of intrinsic microporosity

Taehoon Lee (MIT, ChemE)*; Ho Bum Park (Hanyang University); Zachary Smith (MIT)

"Significant efforts have been devoted to designing novel and high-performance polymers of intrinsic microporosity (PIMs), while little attention has been paid to membrane engineering – converting the developed PIM materials into thin membranes using a procedure appropriate for real-world applications. Herein, we highlight the critical needs for understanding the transition from bulk PIM materials to their thin films to achieve target performance for emerging gas separation applications. To this end, this work will present how to leverage two major challenges in PIM-based thin-film composite (TFC) membranes: (1) plasticization and (2) physical aging.

First, we reveal and quantify the intensified C3 hydrocarbon-induced plasticization behavior in thin-film membranes with decreasing thickness at a sub-micrometer scale. To circumvent this issue, a plasticization-resistant (PR) layer concept was introduced to prevent the C3 hydrocarbon plasticization in PIM-based TFC membranes. Advanced analytical techniques for thin films coupled with mixed-gas transport studies successfully demonstrate the promising applicability of the PR layer for high-performance TFC membranes for C_3H_6/C_3H_8 separation.

Second, we demonstrate that the integration of accelerated aging in PIM thin films and aging-induced selectivity enhancement provides new opportunities to develop high-performance TFC membranes for H_2 purification. To quickly reach an equilibrium state that offers stable separation performance, the aging rate of the as-prepared TFC membrane was hastened by a simple post-treatment step using vacuum storage at high temperature, known as "hyperaging". The resultant hyperaged TFC membranes displayed beneficial aging effects with excellent mixed-gas separation abilities as well as long-term stability for the H_2/CH_4 mixture."

2:00 PM

Oral 182 - CO_2/C_2H_2 Transport in SSZ-13 Zeolite Crystals and Membranes, and the Implication on Zeolite Membrane Microstructure

Ninad Anjikar (Cleveland State University); Kevin Hinkle (University of Dayton); Orhan Talu (Cleveland State University); Qiang Fu (Georgia Institute of Technology); Sankar Nair (Georgia Institute of Technology); Shaowei Yang (Cleveland State University)*

CO_2/C_2H_2 separation is critical for acetylene production. It is highly desirable to achieve the separation with energy-efficient membrane process. Due to their identical kinetic size and close physical properties, the separation is considered very challenging. However, we found for the first time that highly selective CO_2/C_2H_2 could be achieved with SSZ-13 zeolite membranes. Both unary permeation and binary mixture permeation data demonstrated consistent high CO_2 permselectivity > 30. Adsorption and transport properties of CO_2 and C_2H_2 molecules in SSZ-13 zeolite crystals and/or membranes were investigated to understand the origin of selectivity. Equilibrium adsorption isotherms indicate only a slight adsorption selectivity for C_2H_2, implying that a high diffusion selectivity for CO_2 is required for the effective separation based on the solution-diffusion model. The dynamic uptake data, however, points to slight C_2H_2 diffusion selectivity in CHA zeolite crystals. The results indicate that there might be a delicate microstructure in polycrystalline zeolite membranes enabling highly selective separation towards closely sized molecules like CO2/C2H2.

2:20 PM

Oral 183 - Gas Permeation and Separation Characteristics of Microporous TpHz COF Membranes Synthesized by Substrate-Assisted Interfacial Polymerization

Jose E Lopez-Cazares (Arizona State University)*; Jerry Lin (Arizona State University); Kailong Jin (Arizona State University); Kyungtae Kim (Los Alamos National Lab)

Microporous two-dimensional covalent organic framework (2D COF) membranes offer promise for gas separation applications, but their gas transport mechanism remains unclear. In this study, TpHz 2D COF membrane supported on a macroporous nylon substrate is prepared by substrate-assisted interfacial polymerization under mild conditions. Formation of a continuous and dense thin (~300-nm-thick) TpHz layer is confirmed by scanning electron microscopy and Fourier transform infrared spectroscopy. Characterization by X-ray diffraction, grazing incidence wide-angle X-ray scattering, and N 2 porosimetry reveals qualitatively the microstructures of the supported TpHz membranes, i.e., they comprise partially oriented 2D COF lamellar crystallites with moderate crystallinity in an eclipsed (AA) stacking geometry, centering the effective membrane pore size distribution at ~1.1 nm. Single gas permeation data show that the transport of common molecular gases, including H 2 , He, CH 4 , N 2 , and CO 2 , through the synthesized TpHz membranes follows the Knudsen transport mechanism, where single gas permeance decreases with increasing molecular size and permeation temperature. Binary gas separation results show that in the equimolar CO 2 /N 2 mixture, the presence of CO 2 surface flow slightly hinders the N2 flow at room temperature due to the reduced membrane channel size by the adsorbed CO 2 gas layer on TpHz's pore wall. In contrast, permeation of the equimolar CH 4 /N 2 binary mixture does not exhibit discernible surface flow of both gases due to their much lower gas uptake on TpHz, and their transport mechanism follows a Knudsen-like behavior.

2:40 PM

Oral 184 - Towards the Synthesis of Mixed Matrix Membranes Using Vapor Phase Infiltration

Seancarlos Gonzalez (University of Washington)*; Sarah West (University of Washington); Yuri Choe (University of Washington); Kellen McKinney (University of Washington); Cecilia Osburn (University of Washington); David S Bergsman (University of Washington)

Replacing cryogenic distillation and liquid absorption with membranes for gas separations would help decarbonize energy-intensive industries that rely on carbon-based energy and reduce the energy cost of carbon capture. Polymeric gas separation membranes have high processibility and low energy usage but have insufficiently high permeability and selectivity to incentive their use. Mixed matrix membranes (MMMs), composed of a continuous polymer phase and a dispersed phase, have shown improved relative separation properties. However, MMM synthesis typically involves mixing a polymer solution with a solution of dispersed metal-organic frameworks (MOFs). This can introduce interfacial compatibility issues between the polymer and MOF that negatively influence membrane performance. One possible alternative is to use vapor phase infiltration (VPI) to synthesize MMMs. In VPI, a polymer substrate is sequentially exposed to vapor-phase reactants that diffuse into and react within the polymer. The incorporated materials can then influence membrane properties, such as thermal and chemical stability, permeability, and selectivity between gas pairs. If metal oxides embedded through VPI could be replaced with MOFs, VPI could be used to convert existing polymers into MMMs as an additional drop-in step to polymer membrane manufacturing.
Here, we show our progress towards the synthesis of MMMs using VPI. First, we report the infiltration of zinc oxide and 2-methylimidazole, the reactants needed to form the MOF known as ZIF-8, using depth-profiled XPS and SIMS. We demonstrate the deposition of ZIF-8 thin films using ellipsometry and XRD. Glow-discharge optical emission spectroscopy (GD-OES) was used in conjunction with XRD to observe any crystallinity within the polymer interior. Lastly, we will discuss our efforts to characterize

the nucleation and growth of ZIF-8 inside bulk polymers. This approach, when successful, could enable the scalable manufacturing of MMMs.

Solute Selective Separations Modeling

1:00 PM

Oral 185 - Ion and water dynamics in the transition from dry to wet conditions in salt-doped PEG

Nico Marioni (The University of Texas at Austin)*; Oscar Nordness (University of California Santa Barbara); Zidan Zhang (The University of Texas at Austin); Rahul Sujanani (University of California Santa Barbara); Benny Freeman (The University of Texas at Austin); Rachel Segalman (University of California Santa Barbara); Raphaële Clément (University of California Santa Barbara); Venkat Ganesan (The University of Texas at Austin)

Ion transport properties are widely studied in polymer electrolytes under rigorously dry conditions, e.g., in energy storage applications, and in highly swollen hydrogel networks, e.g., in water purification applications. While our knowledge of ion transport in the extreme cases of dry and highly swollen polymer membranes is relatively well developed, the transition between these extremes, i.e., the low hydration regime, is poorly understood. In this study, we combine ion and water diffusivity (PFG-NMR) measurements with atomistic molecular dynamics simulations to better understand transport processes in hydrated salt-doped poly(ethylene glycol). Above the water percolation threshold, hindered and free volume transport models are in good agreement with experimental and simulated diffusivities. At low hydration levels, unlike dry systems, ion diffusion cannot be described by polymer segmental dynamics alone. We rationalize such observations by the interplay between ion-water and ion-polymer solvation of cations, and between ion-water and cation-anion interactions for anions. Further, we demonstrate that a two-state model combining ion-water solvation and free volume transport can describe water dynamics across the entire hydration range of interest. Our findings provide a more encompassing analysis of ion and water transport in hydrated polyelectrolytes, specifically in the low hydration regime.

1:20 PM

Oral 186 - The influence of Donnan – Manning and dielectric exclusion on ion partitioning in sulfonated polysulfone

Sean M Bannon (University of Virginia)*; Geoffrey M Geise (University of Virginia)

"The Donnan – Manning model is the current state-of-the-art approach used to describe thermodynamic properties (e.g., ion sorption) in charged polymer membranes. While the Donnan – Manning model accurately describes the thermodynamic properties of many charged polymers equilibrated with aqueous electrolytes, such as the commercial ion exchange membranes CR61 and Nafion, it less accurately describes the ion sorption properties of sulfonated polysulfone, which is a promising desalination membrane polymer. One possible explanation is that other ionic interactions that are not included in the Donnan – Manning model, e.g., dielectric exclusion (or ion solvation effects) that can be described via the Born model, may have a significant influence on ion sorption in sulfonated polysulfone.

This presentation discusses (1) a modelling approach to account for the influence of both Donnan – Manning and dielectric exclusion factors on ion sorption properties of charged polymers and (2) the application of this model to sulfonated polysulfone and Nafion. The reported Donnan – Manning – Born model describes the ion sorption properties of sulfonated polysulfones more accurately than the Donnan – Manning model alone. The Donnan – Manning – Born model also predicts ion sorption properties in Nafion with improved accuracy relative to Donnan – Manning predictions, but the improvement is less significant than in the sulfonated polysulfone case. These results provide an approach for quantifying the influence of dielectric exclusion on ion sorption in charged polymers and could be particularly important for materials where dielectric exclusion is relevant."

1:40 PM
Oral 187 - Influence of backbone rigidity and water content on lithium/magnesium selectivity in hydrated polymer membranes

Paul R Irving (University of Texas at Austin)*; Nico Marioni (The University of Texas at Austin); Harnoor Sachar (University of Texas at Austin); Everett S Zofchak (The University of Texas at Austin); Kevin Reimund (University of Texas at Austin); Zidan Zhang (University of Texas at Austin); Tyler J Duncan (The University of Texas at Austin); Benny Freeman (The University of Texas at Austin); Venkat Ganesan (University of Texas at Austin)

Increasing lithium demand requires new and improved methods for purification. Membrane-based separation processes, such as selective electrodialysis, hold promise for efficiently separating lithium from other contaminants, such as magnesium. Successful implementation of these technologies requires materials with high selectivity of lithium over other ions while maintaining high lithium conductivity. Two factors that are known to play important roles in governing membrane performance are equilibrium water content and polymer segmental dynamics. We perform both experiments and coarse-grained molecular dynamics simulations to examine the interplay between these two factors and their impact on lithium/magnesium selectivity.

2:00 PM
Oral 188 - Molecular Dynamics Simulations of Lithium/Magnesium Mixed Solutions in Polyamide Membranes

Richard Lueptow (Northwestern University)*; Sinan Keten (Northwestern University); Suwei Liu (Northwestern University)

Nanofiltration (NF) membranes offer cost-effective solutions to ion separations for lithium recovery. However, the effect of feed pH on ion selectivity at various solution concentrations is not well understood. While size exclusion plays an important role in solute rejection, membrane charge interactions with mixed solutes can also influence the selectivity of mono-/divalent ions for polyamide membranes. To better understand solute-solute and solute-membrane interactions, molecular dynamics simulations are used to model the filtration processes. Previous MD studies focus on single solutes at only one concentration, simulated at only one feed pH. In this study, we consider several feed pH values in the form of membrane charges (ionized amine and carboxylate groups). Additionally, we simulate mixed lithium/magnesium feed solutions at different concentrations. This study provides molecular insights with the broader aim of improving the design of NF membranes for solute selective separations.

2:20 PM
Oral 189 - Modeling polymer aging in solvent environments with symbolic regression

Matthew Rivera (MIT)*; Zachary Smith (MIT)

Polymeric membrane separations present an energy-efficient alternative to established, thermally-intensive separation processes. The majority of membrane polymers are glassy materials, which enable high selectivities, but their performance also changes over time due to conformational changes in the polymer microstructure towards a lower-energy state. These changes present a barrier to widespread membrane adoption due to uncertainty about long-term performance stability. Despite its practical importance, glassy polymer aging behavior is still a poorly understood phenomenon, especially in chemically harsh solvent environments. Here, we attempt to understand the relationship between operating environment and polymer aging behavior in solvents using symbolic regression. Symbolic regression is a machine learning paradigm wherein data is fit to a function, as in normal linear regression, but the algorithm additionally finds the most appropriate functional form for the provided descriptors. This strictly empirical approach has the benefit of uncovering non-obvious relationships that might otherwise be overlooked in physics-focused derivations. The inputs to the algorithm are solvent physical properties like kinetic diameter, solubility parameters, etc. We use high-resolution permeance vs. time data to develop models that predict the aging rate, the decline in flux due to aging, and the steady state permeance for Matrimid thin film composite membranes across a range of aromatic, non-polar, and polar solvents. We find that descriptors related to solvent condensability heavily influence the aging behavior of the polymer and that the models can predict Matrimid solvent aging behavior with high accuracy. This case study of a common polymer furthers our understanding of glassy polymer aging behavior in solvent-based membrane separations and highlights the ability of machine learning to provide quantitative structure-property relationships.

2:40 PM

Oral 190 - Ion-ion Separations in Biomimetic Water Channels

Tyler J Duncan (The University of Texas at Austin)*; Harekrushna Behera (The University of Texas at Austin); Nico Marioni (The University of Texas at Austin); Meron Tadesse (University of Texas Austin); Paul R Irving (University of Texas at Austin); Everett S Zofchak (The University of Texas at Austin); Manish Kumar (University of Texas at Austin); Venkat Ganesan (University of Texas Austin)

Demand for critical elements has increased exponentially as the implementation of batteries (Li) and solid-state devices (Nd, Tb, & Eu) have become ubiquitous. A key limiting step in the extraction of these critical elements is understanding ion-membrane interactions that result in high permselectivies. Through molecular dynamics simulations, we analyze an energy efficient and environmentally beneficial membranes for ion-ion separations that are embedded with artificial water channels. Such channels display angstrom-scale separations and tunable ion-ligand binding energies that mimic transmembrane proteins. Using the ligand-appended pillarene chemistry, we have implemented a range of techniques to evaluate ion transport and correlated the effects of channel chemistry to ion permselectivity. The combination of these techniques bridge monomer-ion interactions, which can be resolved on a very fast timescales, to ion-channel interactions that must be resolved through free-energy sampling methodologies due to larger timescales. Using these techniques, we investigate the ion selectivity of three ligand chemistries: diphenyl-N,N-diisobutyl carbomoyl phosphine oxide, diglycolamine, and propionamide phosphonic acid. These chemistries have displayed novel liquid-liquid separation potentials for medium and heavy lanthanide species. Comparing the chemical structures and the resulting permselectivity calculations, we will provide insight into how the placement of hydrophilic groups along appended-ligands influence the selectivity and permeability of our channels.

NAMS Keynote II

1:00 PM

Oral 191 - Dirt and waste as source materials for resource recovery membranes

Seth Darling (Argonne National Laboratory)*

Surface water, groundwater, and wastewater are emerging as potential new sources to bolster supply chains of critical resources. Implementing sustainable—and ideally circular—pathways to recover these resources is an essential strategy to ensure secure supply chains. Membrane separations can offer energy efficiency, but the membranes themselves can also benefit from consideration of circularity in their manufacturing. We report an approach to tuning transport channels in membranes comprised of naturally sourced materials (e.g., exfoliated and restacked vermiculite or montmorillonite, Figure 1) with molecular cross-linkers to control the interlayer spacing, enhance the membrane stability, and manipulate the chemical and electrostatic environment in the channels. The as-prepared cross-linked 2D phyllosilicate membranes exhibit ion diffusivities tuned by the length of the selected cross-linking molecule. A second method for membrane-based resource recovery is through the use of photothermally enhanced evaporation. Evaporation ponds are widely used in resource recovery operations and can often represent a bottleneck in production. Accelerating the evaporation rate using membranes integrating biochar and other waste materials can significantly and sustainably increase supplies of critical materials such as lithium and phosphate.

1:30 PM

Oral 192 - Pall's Filtration Solution for Separation and Purification in Enabling Decarbonization

Varun Vakharia (Pall Corporation)*

Filtration technology has evolved from a straightforward sieve process to cutting-edge functionalized-interactive system for targeted removal. Pall's unique approach to media design and membrane technologies (for separation and purification) drives innovation, especially in the decarbonization space (energy transition mission). Pall's filtration solutions play unique role in the decarbonization space across traditional carbon capture process/refineries to membrane-based CO2-removal/hydrogen purification technologies. Pall is a global leader is delivering customized separators/coalescer solutions for multi-phase system across refineries, chemical industry, electrolyzer and fuel cell. Additional, Pall's extensive media capabilities (for gas separation, liquid purification and ion conductivity) plays a critical role in enabling the electrification/fuel-cells/carbon-free liquid fuel/membrane-based carbon-capture technologies. This integrated approach (via technology adoption and collaboration) across the entire energy space is expected to play a critical role (with differentiated impact) in delivering the net zero targets for 2050.

2:00 PM

Oral 193 - Opportunities and challenges for CO2 separation membranes in decarbonization of the electric power, cement, and steel industries

David Hopkinson (NETL)*; Lingxiang Zhu (National Energy Technology Laboratory); Thien Tran (National Energy Technology Laboratory); Victor A Kusuma (NETL); Eric Grol (NETL); Timothy Fout (NETL); Grigorios Panagakos (Carnegie Mellon University); Glenn Lipscomb (University of Toledo, Toledo); Brenda Petrilena (U. S. Steel); Neil Pergar (U. S. Steel)

According to US EPA, 6.3B tonnes of CO2 were emitted within the US in 2021 with 25% attributed to electric power and 23% from industry. While the US strives for net zero emissions by 2050, the abatement of CO2 from these sectors continues to progress slowly. This is a testament to the magnitude of the challenge, as well as the opportunity for carbon reducing technologies like membrane-based CO2 capture. The US DOE has sponsored membrane R&D for two decades through the Carbon Capture program. Here we present an overview of progress and challenges for this expansive body of work.

This talk will first cover notable DOE projects for carbon capture. CO2 capture from coal-fired power plants was the earliest application studied for membranes. DOE has funded demonstrations at the National Carbon Capture Center since 2008. One challenge with CO2 capture from coal-fired flue gas is its low partial pressure driving force of CO2 (~0.10–0.14 bar). As coal-fired power declines in the US, focus for CO2 capture has shifted to industrial applications including cement and steel production. Demonstrations from MTR for cement and from NETL for steel are underway. These applications may be more favorable for gas separation membranes because of higher partial pressure of CO2 (~0.2–0.3 bar). However, many challenges remain. As an example, false air ingress in cement plants lowers the effective partial pressure for CO2. In steel production, contaminants may poison membranes.

We will present NETL's economic analyses for membrane-based CO2 capture from coal-fired flue gas and from cement. Compared to the traditional amine solvent approach, membranes feature lower energy consumption and simplicity of operation. However, in some scenarios the cost of capture for membranes can be favorable compared with amines, but in others it is not.

Last, we will discuss NETL's membrane research towards higher performing membrane materials, more efficient membrane module design, and field demonstrations.

2:30 PM

Oral 194 - Carbon capture through membranes: leveraging multiphysics modeling, dimensional analysis and machine learning to scale up and optimize devices and processes for decarbonization

Hector A Pedrozo (Carnegie Mellon University); Cheick Dosso (Carnegie Mellon University); Thien Tran (National Energy Technology Laboratory); Lingxiang Zhu (National Energy Technology Laboratory); Victor A Kusuma (NETL); David Hopkinson (NETL); Lorenz Biegler (Carnegie Mellon University); Grigorios Panagakos (Carnegie Mellon University)*

The operational efficiency of membrane modules depends on several independent and interrelated process variables. As the process scale transitions from lab scale to industrial applications, predicting the performance of the separation becomes crucial. Our study addresses this challenge by using dimensional analysis (DA), a well-established technique for designing full-scale processes based on exploiting lab experiments and simulations across different scales or process conditions. Through DA, dimensionless numbers are generated from dimensional variables to describe the physics of the processes. This approach leads to simplified correlations, providing insights into the impact of different scaling parameters. Furthermore, these correlations can be use as tools for performance prediction in operational regimes where getting experimental data or computational simulations is limited by resources.

As a case study, we examine the separation of CO2 from a binary mixture with N2 using membrane modules, which is a relevant scientific problem for carbon capture research. To develop the correlations, we use both experimental data and simulations performed with a validated computational fluid dynamics (CFD) model. The CFD models offer valuable insights, covering scales and process conditions not included in lab experiments.

While well-known dimensionless numbers associated with transport phenomena, including the Reynolds, Peclet, Schmidt, and flux factor, were considered, our study suggests that the strongest correlations in membrane modules are found when using dimensionless numbers specifically tailored to membrane separations. Even increasing the membrane module scale nine times, we find that the dimensionless feed flow number can serve as an input variable to determine the stage cut of the membrane module, resulting in an R2 correlation coefficient exceeding 0.99. Beyond this case study, the methodology could be applied to gas separations for different membrane modules.

Contaminants Removal from Water II

1:00 PM

Oral 195 - Real-Time Induced Magnetic Vibrational Based Antifouling Mechanism for Ultrafiltration (UF) membrane

Jasneet K Pala (The University of Alabama)*; Milad Rabbani Esfahani (The University of Alabama); Ryan Tracy (The University of Alabama); Nima Mahmoodi (The University of Alabama)

Despite the advanced capabilities of membrane technologies, which can produce billions of gallons of water daily, fouling remains a significant challenge. Fouling reduces membrane separation efficiency by compromising permeability and rejection, thereby shortening membrane lifespan and escalating process costs through heightened washing requirements. Chemical modification of membranes, while promising, may lead to drawbacks like reduced stability, selectivity, and flux due to structural changes. In this study, we report a physical and in-situ antifouling approach by investigating the effects of real-time induced magnetic vibrations on the membrane surface to hinder or delay the deposition of foulants. In our previous research, we demonstrated that attaching iron particles to the backside of the membrane and applying vibrations via an external magnet at 5 Hz, during treatment with 45 ppm of humic acid (HA), led to a significant reduction in fouling. This resulted in only a 10 % decline in flux over 6 hours of filtration, compared to a substantial 33.9 % flux reduction observed over 6 hours with no vibrations for the blank membrane. In continuation with the previous work, we focused on chemically synthesizing magnetic membranes by incorporating Fe3O4 nanoparticles on the backside of the membrane using the spin coating technique to get a uniform distribution of iron nanoparticles across the membrane, unlike mixed matrix membrane (MMM). The primary emphasis of this work is on synthesizing a magnetic membrane utilizing a spin coater with a two-layer coating. The bottom layer is comprised of Fe3O4-PSF, serving as the magnetic layer, while the top layer consists of a 15% PSF membrane. The performance of the magnetic membranes was investigated under a magnetic field, comparing water flux, rejection, and antifouling properties with blank membranes and mixed-matrix membranes. This evaluation aimed to assess the effects of iron nanoparticle orientation on membrane vibrations.

1:20 PM

Oral 196 - Catalytic Membranes Based on Reduced Graphene Oxide for Micropollutant Treatment Enabled by Peracetic Acid

Erda Deng (State University of New York at Buffalo)*; Siavash Mohamadi (State University of New York at Buffalo); Zachary Kralles (State University of New York at Buffalo); Ning Dai (University at Buffalo); Haiqing Lin (The State University of New York at Buffalo)

Peracetic acid (PAA) has emerged as an interesting oxidant to degrade micropollutants (MPs) with high toxicity in wastewater, and it needs to be activated by carbon materials to generate radicals for decontamination. Herein, we demonstrate, for the first time, that PAA can be combined with reduced graphene oxide (rGO) membranes to degrade MPs, such as 2,4-dichlorophenoxyacetic acid (2,4-D) and methylene blue (MLB). First, rGO membranes with various thicknesses (100 – 500 nm) were successfully prepared and evaluated in a crossflow system. the 2,4-D content remained unchanged in the PAA solution, but when the 2,4-D and PAA mixture was filtered through the rGO membrane, 30% degradation of 2,4-D was achieved. This synergy was further highlighted in the degradation of MLB, where the rGO/PAA catalytic membrane system exhibited significant efficacy compared to using PAA alone. With a PAA content of 25 ppm in the feed, the degradation of MLB is over 30%, which is higher compared to only 5% degradation without PAA. Interestingly, the 100-nm rGO membranes displayed a six-fold increase in water permeance compared to the 500-nm counterpart, yet both membranes exhibited similar MLB degradation rates. Importantly, our findings indicate that the catalytic property remains independent of rGO thickness, as evidenced by the consistent performance described by Pseudo first-order kinetic. Remarkably, the 100-nm rGO-modified membrane used to purify wastewater containing MLB in a crossflow system exhibits a degradation of MLB of more than 3 g/(m2·h) with 25 ppm PAA.

1:40 PM

Oral 197 - Designing the fluorinated superfine-activated carbon adsorptive membrane for selective removal of long- and short-chain PFAS

Medha Kasula (The University of Alabama)*; Milad Rabbani Esfahani (The University of Alabama)

Membrane filtration and adsorption are two prominent methods for the effective removal of per- and poly-fluoroalkyl substances (PFAS) from water. Thin-film composite (TFC) nanofiltration (NF) membranes exhibit acceptable PFAS removal (<90%), but the presence of rejected PFAS in the concentrate stream necessitates an additional separation process. Our previous study reported the fabrication of adsorptive TFN NF membranes, incorporating superfine activated carbon (SFAC), for enhanced perfluorooctanesulfonic acid (PFOS) removal via adsorption during water filtration. The SFAC-TFN-0.5 membranes (0.5 wt% loading) of SFAC embedded inside the polyamide layer of TFN demonstrated 94% PFOS removal, with over 80% adsorption, and an increased permeability of 22 LMH/bar compared to the bare TFC. Although the results were impressive, the selectivity of the membranes towards different PFAS (long-chain PFOS and short-chain perfluorobutane sulfonate, PFBS), and in the presence of other components was not investigated. In this project, we investigated the selectivity of the adsorptive membrane toward different PFAS in the presence of NOM. Initial static adsorption results revealed limited PFBS adsorption on SFAC (45%), while it showed 93 % adsorption towards PFOS. To enhance the adsorbent's selectivity for different PFAS, SFAC was further fluorinated, as fluorination increases fluorophilic interactions with short-chain PFAS — an essential mechanism for their adsorption. The selectivity of PFOS in the presence of NOM on the SFAC membrane was observed to decrease to 70 % from 90 %. Therefore, the hypothesis is that fluorophilic interaction can boost the affinity of SFAC towards PFOS rather than NOM. In addition, the incorporation of SFAC in the IP process led to SFAC aggregation and disruption of the polyamide layer as reported in our previous study. In this project, fluorinated SFAC is grafted onto the active layer to prevent aggregation and better control surface properties.

2:00 PM

Oral 198 - Dynamic ammonium retention for nutrient separation from manure digestate

Marrit van der Wal (Eindhoven University of Technology)*; Zandrie Borneman (Eindhoven University of Technology); Kitty Nijmeijer (Eijndhoven)

The agricultural sector wants to further valorize manure by producing separate nutrient-rich fractions which allow to tailor the nutrient ratio per crop and season for precision fertilization, reducing the need for artificial fertilizers. The existing manure treatment often involves the production of biogas by co-digestion of the manure and subsequent mechanical and membrane separation steps resulting in a liquid fraction containing nitrogen and potassium. Separating these nutrients from the liquid fraction has proven particularly difficult as the nitrogen is mainly present in the form of ammonium ions (NH4+) which has a similar charge valency, ionic radius and hydration shell as potassium ions. However, the acid-base equilibrium of NH4+ provides the opportunity for separation as increasing pH shifts the NH3/NH4+ equilibrium toward neutral NH3 which permeates through the membrane more easily. At a high pH, the K+ is mostly retained, whereas the NH3 permeates through the membrane separating the nutrients. In this research, we investigate the separation of nitrogen and potassium for nanofiltration (NF) and reverse osmosis (RO) membranes in a dead-end cell and a crossflow system by varying the pH of the artificial NH4+/ K+ mixtures and real digestate. No nutrient separation was observed at neutral pH conditions for either the artificial single salt solutions or the mixture. However, significant Total Ammonium Nitrogen (TAN)/K selectivity was achieved at alkaline feed solution conditions for both the artificial mixture and the real digestate for the RO membranes. With a dynamic pH approach in a multistep RO system, the production of a separate nitrogen-rich stream can be achieved. Combining an alkaline RO step (separation) followed by an acidic RO step (concentration) of the permeate stream results in the required nitrogen/potassium separation followed by the nitrogen concentration.

2:20 PM

Oral 199 - Advanced functionalized polyether ether ketone/polysulfone membranes with 2D nanomaterials for enhanced water treatment performance

Muhammad Usman Yousaf (University of Kentucky)*; Nerissa Kiprotich (University of Kentucky); Anika Azme (University of Nebraska, Lincoln); Lucca Madeo Cortarelli (University of Kentucky); Nirupam Aich (University of Nebraska, Lincoln); Olga Tsyusko (University of Kentucky); Isabel Escobar (University of Kentucky)

Nanofiltration (NF) is capable of selectively removing contaminants, low energy consumption and versatility for a variety of water treatment applications. However, the removal of emerging contaminants, such as organic chemicals, pharmaceuticals and micropollutant, remains a challenge for the NF technology. There is a need to design NF membranes with specialized surface modifications and pore sizes to target specific emerging contaminants. Two-dimensional (2D) nanomaterial-based NF membranes present an effective approach in water purification due to their confined nanocapillaries, and electronic and optical properties imparting superior performance in terms of water flux, rejection rates, and resistance to fouling . Non-carbon 2D photocatalysts like phosphorene and hexagonal boron nitride (h-BN) offer promising capabilities for water purification by combining separation with the photocatalytic destruction of pollutants, including fluorinated chemicals, through light-induced reactive oxygen species

(ROS). Phosphorene, known for its unique electronic properties and high carrier mobility, and h-BN got strong sp2 bonded honeycomb structure, degrading pollutants due to its ability to generate electron-hole pairs under UV light. Herein, 2D materials-based photocatalysts were incorporated into sulfonated polyether ether ketone/polysulfone (SPEEK/PSF) composite membrane matrixes to introduce advanced catalytic functionality to the membrane. The SPEEK polymer in the composite membrane provides ion-exchange properties or water uptake ability, while the PSF imparts mechanical strength to enhance the overall stability of the membrane. The combination of these two materials with the functionalization of 2D nanomaterials results in a multifunctional membrane capable of removing and degrading organic contaminants simultaneously. The synthesized membranes undergo comprehensive surface and structural characterizations, including evaluations of mechanical and chemical stability.

2:40 PM

Oral 200 - Distributed Membrane-Based Water Desalination and Purification Systems: Automation, Data-Driven Operational Models, and Performance Forecasting

Nora Marki (University of California, Los Angeles) *, Yoram Cohen (University of California, Los Angeles), Yang Zhou (Shanghai University), Yakubu Jarma (University of California, Los Angeles), Maria Soto (University of California, Los Angeles), Christian Aguilar (University of California, Los Angeles), Phoebe Strauss, Muhammad Bilal (University of California, Los Angeles)

Distributed membrane-based water desalination/treatment (DWTD) systems are suited for industrial facilities, produced water treatment, local upgrading of impaired potable groundwater supplies, agricultural and drainage water. However, effective deployment of DWTD systems requires effective remote monitoring and autonomous operation that responds to water feed quality fluctuations, water demand variability, target product water quality, and minimization of treatment residuals. Accordingly, a novel autonomous membrane-based water desalination/purification system architecture was developed, introducing real-time system performance monitoring, machine-learning (ML) operational models for optimal operation, and AI-driven decision support. ML model forecasting of system performance was developed enabling early detection of system performance degradation (e.g., due to fouling), and identification of sensor faults. To accomplish the above, a unique cyberinfrastructure was developed for managing multiple DWTD systems via real-time system performance monitoring and data analytics, in addition to supervisory control and asset management. Over a three-year operational experience with three wellhead DWTD systems, robust operation was demonstrated owing to advanced data analytics for forecasting performance degradation which was instrumental in avoiding operational upsets. Moreover, an approach developed for predicting membrane performance, based on data-driven process models, enabled forecasting of solute transport flux and extraction of fundamental salt permeability coefficient dependence on operating conditions (including temperature). It is also noted that system operation with partial concentrate recycle was shown to be effective for decreasing system footprint, and that increasing single pass recovery reduced energy consumption. The above DWTD systems now provide safe potable water to three communities in Salinas Valley, California.

NAMS Sponsors - Diamond Level

NSF

MEWBOURNE
COLLEGE OF EARTH AND ENERGY
The UNIVERSITY *of* OKLAHOMA

INSTITUTE FOR RESILIENT
ENVIRONMENTAL AND ENERGY SYSTEMS
The UNIVERSITY *of* OKLAHOMA

NAMS Sponsors - Gold Level

PALL®

NAMS Sponsors - Silver Level

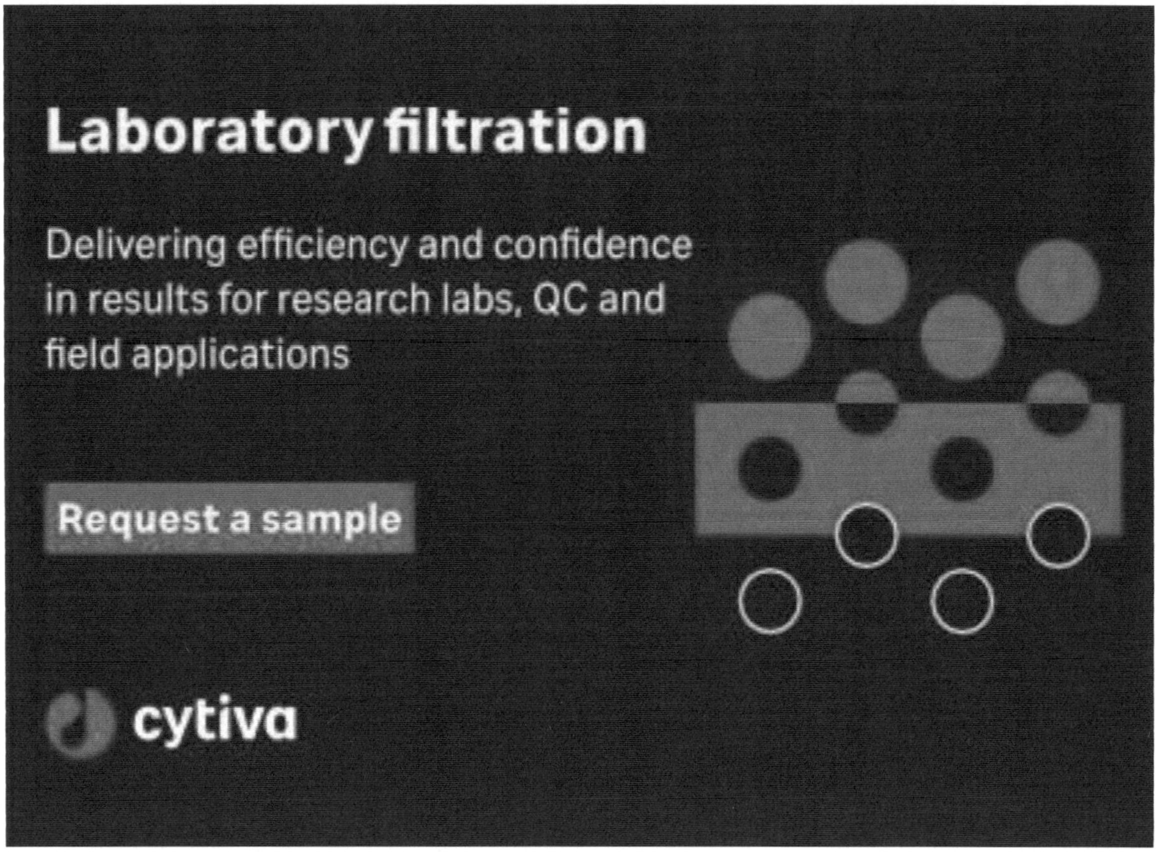

Seek Together™

DOW®

it's only human
to see the
future runs on
innovation
Chevron proudly supports NAMS
the
human
energy
company
Chevron
CHEVRON, the CHEVRON Hallmark and THE HUMAN ENERGY COMPANY are registered
trademarks of Chevron Intellectual Property LLC. © 2023 Chevron U.S.A. Inc.

ZwitterCo

ZWITTERCO
ZWITTERCO
MAKE A SPLASH IN THE WORLD
Play a part in developing groundbreaking solutions at ZwitterCo.
Apply today: www.zwitterco.com/careers

NAMS Sponsors - Bronze Level

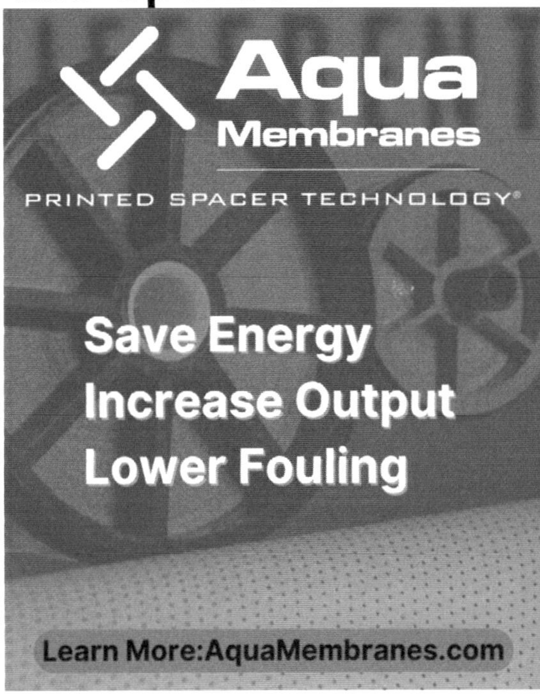

MTR

CARBON
CAPTURE

INDUSTRIAL
SEPARATIONS

MAXWELL
ROBOTICS

Air Liquide

Gale & DB
Bhattacharyya

North American Membrane Society (NAMS)
University of Arkansas
Ralph E. Martin Department of Chemical Engineering
3202 Bell Engineering Center
Fayetteville, AR 72701-1201
USA

ISBN 979-8-3313-0056-2

9 798331 300562

CURRAN ASSOCIATES INC.
proceedings.com